# Coastal and Estuarine Studies

Series Editors:
Malcolm J. Bowman   Christopher N.K. Mooers

*Gyre in Lake Biwa*

*Reproduced with permission from Asahi Shinbun Company*

# Coastal and Estuarine Studies

48

Setsuo Okuda, Jörg Imberger, and
Michio Kumagai (Eds.)

# Physical Processes in a Large Lake: Lake Biwa, Japan



American Geophysical Union

Washington, DC

**Series Editors**
Malcolm J. Bowman
Marine Sciences Research Center, State University of New York
Stony Brook, N.Y. 11794, USA

Christopher N.K. Mooers
Division of Applied Marine Physics
RSMAS/University of Miami
4600 Rickenbacker Cswy.
Miami, FL 33149-1098, USA

**Editors**
Setsuo Okuda
Okayama University of Science
1-1 Ridai-cho
Okayama, Okayama, 700, Japan

Michio Kumagai
Lake Biwa Research Institute
1-10 Uchidehama
Otsu, Shiga, 520, Japan

Jörg Imberger
Centre for Water Research
University of Western Australia
Nedlands, WA 6009, Australia

**Library of Congress Cataloging-in-Publication Data**

Physical Processes in a Large Lake: Lake Biwa, Japan / Setsuo Okuda, Jörg Imberger, and Michio Kumagai, editors.
p. cm. — (Coastal and estuarine studies ; 48)
Includes bibliographical references.
ISBN 0-87590-262-6
1. Limnology—Japan—Biwa Lake. I. Okuda. Setsuo, 1926– .
II. Imberger, Jorg, III. Kumagai, Michio, IV. Series.
GB1752.B58P48 1995
551.48′2′0952185—dc20

95-1870
CIP

**ISSN 0733-9569**

**ISBN 0-87590-262-6**



Printed in the United States of America.

CONTENTS

# PREFACE

This book arose from a discussion between Michio Kumagai, Setsuo Okuda and myself during a visit to Japan under the auspices of the Japan Society for the Promotion of Science. After hearing of the extensive results of the general research on Lake Biwa, it was suggested that much of the work was not well documented in the open literature and could possibly be summarised in book form, complementing the already existing volume "Lake Biwa" by Shoji Horie with a more detailed treatment of the understanding of the physical processes. A further outcome of these discussions was the suggestion of a large interdisciplinary experiment to study the transport processes in Lake Biwa and the impact of these processes on the biogeochemical cycles. This latter suggestion was realised through the Lake Biwa International Transport Experiment (BITEX), the results of which will be published shortly.

The work in Lake Biwa was motivated by concerns about the deteriorating quality of the water. The great majority of pollutants are discharged into the smaller South Basin but these are progressively being transported upstream into the North Basin; the source of drinking water for 15 million people. The urgent management problems have been the motivation for research projects with both short and long term implications. This research should be of interest to managers of water supplies throughout the world.

The authors of this book give a clear exposition of the various transport mechanisms operating in a large stratified lake. The discussion is confined to examples from Lake Biwa, but physical processes such as basin scale gyre formation, water mass exchange, and the role of the benthic boundary layer are relevant to all large lakes. The material presented on these topics does not find a counterpart in open literature and so should be of great interest to limnologists the world over. The chapter on remote sensing techniques is of general interest since it presents for the first time the laser sensing methodology pioneered in Japan.

In this book I have thus attempted to bring together various contributions that detail the advances made in our understanding of the physical processes in Lake Biwa. Also, for the first time, the book provides a coherent overview of the very considerable contributions recently made by Japanese physical limnologists.

I would like to thank the authors for their patience with the long delays in getting the manuscripts into a book form, the three reviewers Greg Ivey, Robert Spigel and Paul Hamblin for their detailed suggestions which led to very considerable improvements in the presentation, and Karen Lane for her very careful preparation of the book as a whole.

Her contributions were not confined to the presentation, but also ranged over the scientific consistency of the manuscripts. Without her input the book would not have been completed. The collaboration and friendship extended to me by Michio Kumagai and Setsuo Okuda during the editing stage made this undertaking a great pleasure; Michio Kumagai must take all the credit for interacting with various authors of the book.


Jörg Imberger
Centre for Water Research
University of Western Australia
Nedlands, WA 6009, Australia

# 1

# Introduction

S. Okuda and M. Kumagai

Lake Biwa is very important in a social context. It is the largest lake in Japan and presently supplies some 15 million people in four prefectures, Shiga, Kyoto, Osaka and Hyogo with drinking water, making it the single most important source of water in Japan.

It is also important from a geographical perspective. It is a natural lake born some five million years ago and is therefore geologically one of the oldest lakes in the world .

Lake Biwa occupies one-sixth of Shiga Prefecture, which is located in the middle of Japan's main island, Honshu. The climate is temperate, with some variations from the northern section to the southern section of the lake. The mountains which form the northern border of the lake trap more than 100 cm of snow each winter, as storms come across the Japan Sea from the Siberian plains. The southern portion of the lake receives much less snow, and can be oppressively hot and humid during the summer. There is a rainy season early each summer (usually lasting from mid-June through mid-July) and a second, less severe period of rain and wind storms each September. The monthly average air temperature and rainfall for Hikone, located midway along the lake's eastern shore, are presented below:

TABLE 1.1. Monthly average temperature and rainfall

| | Jan. | Feb. | Mar. | Apr. | May | June | July | Aug. | Sep. | Oct. | Nov. | Dec. | Annual |
|---|---|---|---|---|---|---|---|---|---|---|---|---|---|
| Mean air temp. | 3.2 | 3.5 | 6.2 | 11.7 | 16.6 | 20.8 | 25.1 | 26.4 | 22.3 | 16.4 | 10.8 | 5.9 | 14.1 |
| Precip. (mm) | 110 | 106 | 119 | 154 | 149 | 224 | 235 | 150 | 206 | 117 | 85 | 86 | 1741 |

Physical Processes in a Large Lake: Lake Biwa, Japan
Coastal and Estuarine Studies, Volume 48, Pages 1–6

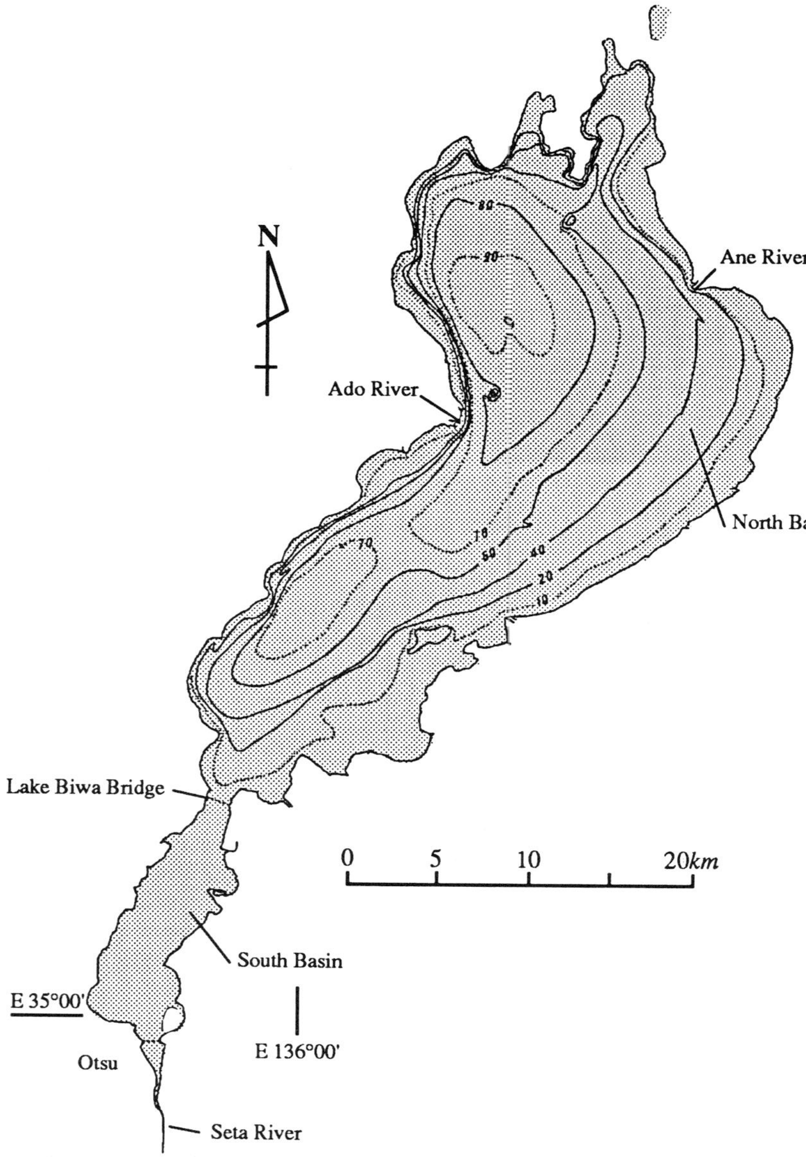


Figure 1.1. Lake Biwa

Lake Biwa is 63.5 km long from north to south and is strongly constricted, reaching a minimum width of only 1.35 km about 16 km from its southern end. The main basin north of the constriction is called the North Basin, while the sub-basin south of the constriction is called the South Basin. The North Basin has an average water depth of 44 m, but the South Basin is only 3.5 m deep on average. Topographic maps of the lake are presented in Figure 1.1. The characteristics of the lake are presented below in Table 1.2:

TABLE 1.2. Lake Characteristics

| | North Basin | South Basin | Total | Unit |
|---|---|---|---|---|
| Altitude | - | - | 85.6 | m |
| Surface area | 616 | 58 | 674 | $km^2$ |
| Volume | $27.3 \times 10^9$ | $0.2 \times 10^9$ | $27.5 \times 10^9$ | $m^3$ |
| Maximum depth | 104 | 8 | 104 | m |
| Mean depth | 44 | 3.5 | 41 | m |
| Period | | | | |
| Seiche | | | 255 (mode one) | min |
| Kelvin waves (Sept.) | | | 60 (mode one) | hour |
| Inertia oscillation | | | 20 | hour |
| Vertical temperature difference (Sept.) | | | 18 | °C |

Lake Biwa's drainage basin is surrounded by mountain ranges about 1,000 m above sea level, and is 4.7 times as wide as the lake itself, corresponding closely to the administrative limits of Shiga Prefecture. Forest-covered hills and mountains account for nearly 60% of the total area of the prefecture (exclusive of Lake Biwa), while farmland, mostly paddy fields, makes up 25% of the land area. Most of the existing forest vegetation is pine forest (*Pinus densiflora*) and mixed deciduous broadleaf forest, both of which were used for firewood and charcoal production until the late 1950s and early 1960s. There are only a few cities of moderate size in the drainage basin, the largest being Otsu with a population of 260,000 (1990). Otsu and the area surrounding the lake's South Basin have recently become popular suburbs of the larger cities of Kyoto and Osaka, and there has been a steady increase in population since the mid-1980s.

Some 460 streams of various sizes flow into Lake Biwa, but the Seta River is the only natural outlet, flowing from the lake's southern end. The Seta is a tributary of the Yodo River, which flows into Osaka Bay (Figure 1.2). In 1906, a dam was built at the uppermost end of the Seta to control both the river's flux and the water level of the lake. If the outflow is too severely limited in times of heavy rainfall, the lake shores are threatened by flooding, whereas Osaka and other densely populated downstream areas may be exposed to flooding when too much water is discharged. This conflict in interest between the residents of Shiga Prefecture and the downstream communities has existed since long before the construction of the dam.

Lake Biwa has provided a number of economic benefits to the community. The lake has been valued since prehistoric times for its abundant fish and shellfish. Until this century, surface traffic on the lake provided an easy alternative to difficult overland haulage in this mountainous region. Recently the lake has been used largely as a recreational site and as a destination for tourists. Lake Biwa has contributed to the development of Osaka as an industrial centre, as the lake is a stable source of water for the Yodo River on which Osaka depends. The revival of the old city of Kyoto was also made possible when a canal, completed in 1890, was built to bring fresh lake water across the mountains.

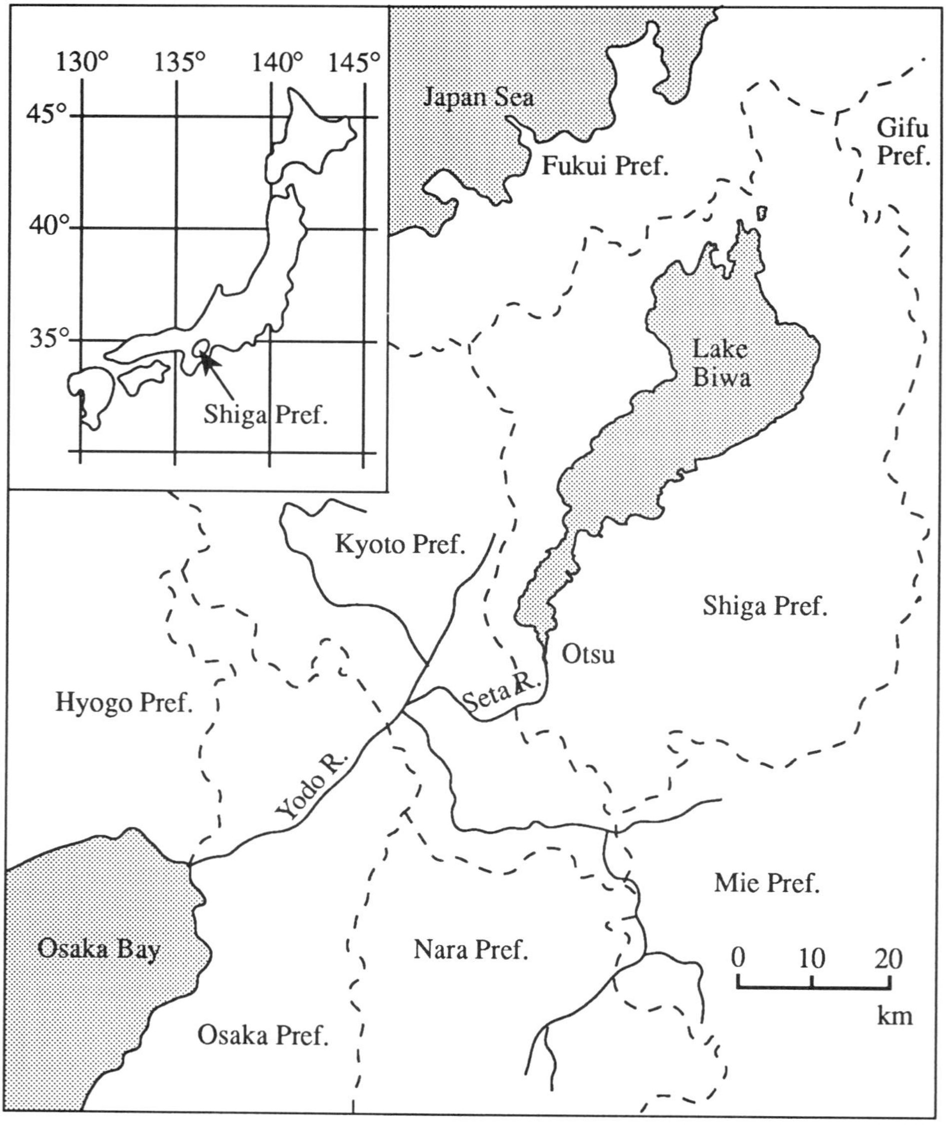


Figure 1.2. Regional location map.

The quality of water in the lake has been profoundly influenced by the economic development of the 1960s and early 1970s, when the trophic level of the Northern Basin changed from oligotrophic to mesotrophic and that of the Southern Basin from mesotrophic to eutrophic. The lake's biota exhibited corresponding drastic changes; tap water drawn from the lake developed a foul smell caused by certain planktonic algae in 1969; and "freshwater red tide" caused by the bloom of a flagellate alga, *Uroglena americana*, first appeared in 1977. These phenomena have recurred every summer since their initial appearances.

Since 1970, the national and local governments have undertaken several projects to prevent and ameliorate pollution of Lake Biwa. In 1970 the national government implemented control of waste water discharge into the lake, slowing eutrophication considerably. In 1980, the Shiga prefectural government enacted an Ordinance for the Prevention of Eutrophication in Lake Biwa, which specified even stricter control of waste water discharge and, for the first time in Japan, prohibited the use of phosphate-containing synthetic detergents.

The Integrated Development Project for Lake Biwa was established in 1972 by the national government and the governments of Shiga and the downstream prefectures. The project is concerned with the development and conservation of the lake's water resources, flood control in the drainage basin and along the Yodo River, and settlement of the conflicts of interest between upstream and downstream districts of the river. Among the project's specific goals are:

- increasing the water supply available to the downstream communities by 40 $m^3 s^{-1}$ to meet increasing demand
- construction of a large-scale sewer system with advanced treatment to prevent further pollution of the lake's water.

A continuous embankment is being built around the lake as part of the effort to limit the effects of fluctuations in lake level.

The Lake Biwa Research Institute was established in April 1982 for the purpose of conserving the natural beauty and water quality of Lake Biwa. Since then, the Lake Biwa Research Institute has conducted fundamental studies with a long-range perspective. The research has been interdisciplinary, covering physical, biogeochemical and social sciences.

This book focuses on the results of physical science research projects conducted in Lake Biwa. Lake Biwa Research Institute has played a pivotal role in most of this research. The research projects encompass six distinct aspects of the physical processes within the lake and the individual papers are arranged accordingly, as follows:

Part I presents an historical review of scientific studies on the current systems in Lake Biwa; starting with the first systematic observations on gyres in 1925.

Part II describes various types of water mass exchange processes which control regional environments.

Part III treats the optical characteristics and turbidity dynamics of the lake. These features are closely related to water quality and biological primary production.

Part IV describes the mixing processes between river water and lake water near the mouth of inflowing rivers.

In Part V, sedimentological and paleo-limnological studies are used to investigate recent environmental changes in Lake Biwa's drainage area.

Part VI introduces sensing and measurement techniques which have been developed during the course of research work in Lake Biwa.

The papers in this book present a variety of studies on different scales. The selection of papers has been restricted to those that mesh together to give a clear picture of the major physical processes in Lake Biwa. Many of the findings are not specific to Lake Biwa, making the book a useful research or teaching tool for limnologists in other parts of the world.

## Reference

Kira, T. (1984) Data Book of World Lakes. A compilation of Physiographic and Socio-Economic Data with Emphasis on Eutrophication. LECS'84.

# 2

# Review of Geostropic Gyres

S. Endoh

There are many kinds of currents in Lake Biwa, having various time scales and dynamics, including wind-driven currents, density currents, inertial oscillations, and currents associated with seiches and internal waves (e.g. Okamoto, 1984). However, the water movement most characteristic of Lake Biwa is a large, well-defined, counter-clockwise cyclonic gyre which is consistently found in the epilimnion during the season of thermal stratification (May through November). Two other gyres, one clockwise and the other counter-clockwise, have also been observed in the lake. In this section, the observational and theoretical studies which have been performed on these gyres are briefly reviewed, and the characteristics of the gyres are described on the basis of the more recent studies.

The Kobe Marine Observatory first carried out systematic observations of Lake Biwa in the summer of 1925. These observations included measurements of currents using Ekman Merz and air-bubble current meters (Suda *et al.*, 1926). This work showed the existence of three gyres in the surface layer of the North Basin and these were named respectively from north to south, the First Gyre (counter-clockwise), the Second Gyre (clockwise) and the Third Gyre (counter-clockwise).

Since World War II, a number of current measurements have been made employing various methods, including the use of Lagrangian drifters. Morikawa and Okamoto (1960, 1962) released a great number of drifter bottles on the lake surface and confirmed the existence of the First and Second Gyres. Okamoto (1968) carried out a series of buoy trackings using cross-board drogues to measure the horizontal and vertical distribution of the water movement. Okumura and Yamamoto (1978) developed a radio tracking buoy system, and Endoh *et al.* (1987) used radar to trace drifters with window-shade drogues over long periods of time, including during the night and under poor

Physical Processes in a Large Lake: Lake Biwa, Japan
Coastal and Estuarine Studies, Volume 48, Pages 7–13

weather conditions. Okuda and Yokoyama (1978) and Obayashi *et al.* (1982) applied aircraft and satellite remote sensing techniques to observe the water movement. Figure 2.1 depicts the locations of the gyres in Lake Biwa obtained by radar tracking of drifters in 1983 and 1986 (Endoh *et al.*, 1987). The third gyre is not shown as it was not clearly observed by radar.

Ekman Merz current meters, operated from a boat, have also been widely deployed (Suda *et al.*, 1926; Morikawa and Okamoto 1962). Automated current meters were introduced and developed in Lake Biwa during the late 1970s (e.g., Okumura, 1977). Endoh *et al.*, (1982) detailed the seasonal variation of the First Gyre by analyzing continuous records of current meters (Aanderaa RCM-4) taken at four stations. A series of continuous current measurements using automated current meters have been carried out successively since 1981 to reveal the spatial distribution and time variations of the current field in Lake Biwa (Okumura and Endoh, 1985; Endoh and Okumura, 1989). These measurements clarified the horizontal and vertical distributions of the gyre motions, the associated periodic current fluctuations, the response of the current to the wind, the characteristics of the current in the deeper layers and the current field in winter.

Okamoto and Morikawa (1961a) analyzed current and temperature data, and found that the gyres in Lake Biwa were in geostrophic balance. More recently, a number of observations on the water temperature field were made using thermistor and bathythermographs (Okamoto and Morikawa, 1961b; Okamoto, 1962, 1968; Kunishi *et al.*, 1967; Imawaki *et al.*, 1979; Endoh *et al.*, 1979, 1981). These observations demonstrated the existence of large amplitude fluctuations in water temperature which may be attributed to internal waves. To improve the geostrophic calculations, it was necessary to obtain time series of the water temperature and then use the time-averaged temperatures. Okamoto (1968) and Endoh and Okamoto (1981) established the validity of the geostrophic balance for the gyre current using simultaneous observation of currents and water temperatures. The results show that the geostrophic approximations were poor when comparing currents observed at a specified time. However, by averaging the current over time so as to cancel the effect of the slowest internal wave signals, such methods yielded serviceable approximations for baroclinic geostrophic gyres.

Endoh (1978) developed a diagnostic model to describe the current field more accurately using the observed water temperature field, and showed the existence of a large vertical circulation in the cyclonic gyre. Using this model, Endoh (1986) evaluated the seasonal variation of the mechanical (kinetic and available potential) energies in the First Gyre and discussed the formation and maintenance mechanisms of the gyres.

Numerical calculations have been performed to evaluate the formation mechanisms of the gyres. Imasato *et al.* (1975) examined the wind-induced barotropic circulation and time variation of the lake level by integrating the linear equations, and found that a geostrophic counter-clockwise circulation could be sustained by a southerly wind. They also evaluated the structure and periods of the surface seiches driven by the wind. Oonishi and Imasato (1975) calculated the time variations of the barotropic current field driven by the wind using a rigid-lid assumption. They found that a depth-integrated gyre could be set up by a steady wind with a topographic curl. They also evaluated the behavior of a topographic Rossby wave with a period of about one week.

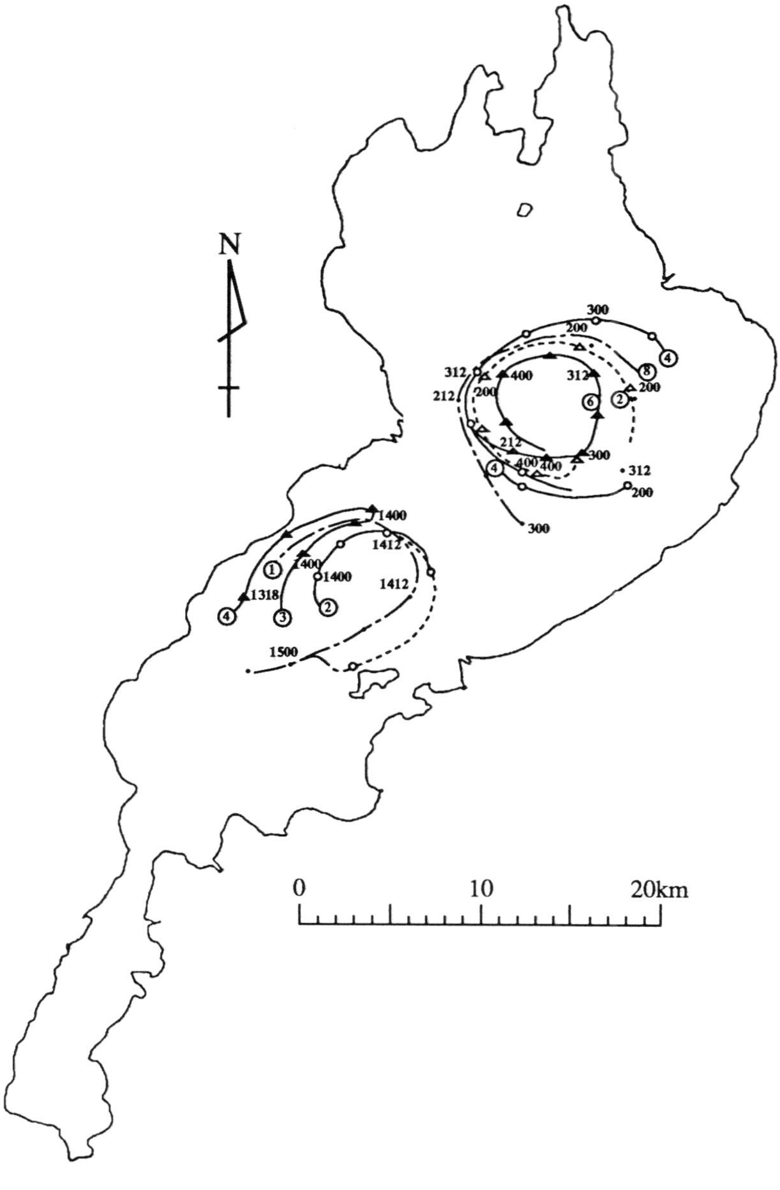


Figure 2.1. The surface gyre currents in the North Basin of Lake Biwa observed by radar tracking of drifters in 1983 and 1986 (after Endoh *et al.*, 1987). Numerals show day and hour (e.g. 312 means 1200 noon UT on 3rd September).

Two-layer models were developed to examine the wind-driven baroclinic current structure and the internal wave shear. Kanari (1974) calculated the structures of various modes of internal waves, especially the internal Kelvin wave. Kanari (1975) then carried out successive observations of the water temperature field in the summer and confirmed the structure of the internal Kelvin wave with a period of about two days. Endoh (1976) then used a two-layer model which included the topography of the lake bottom and demonstrated the possible formation of long-lived gyres supported by the wind stress curl.

The possibility of gyre formation by solar radiation (thermally driven circulation) has also been examined. Oonishi (1975) carried out numerical calculations using an axis-symmetric lake model and was able to simulate the formation and maintenance of a cyclonic gyre (the First Gyre). Ookubo *et al.* (1984) carried out a series of physical laboratory experiments to evaluate the topographic effect on heat accumulation. Itakura *et al.* (1990) measured the time-variation of the surface water temperature in the laboratory model using infrared thermometry. These studies are described in detail in the following section and chapter.

Cyclonic surface circulation has also been observed in other large lakes (Emery and Csanady, 1973). Many theories have been proposed to explain the formation of the gyres (*e.g.* Imberger and Hamblin, 1982). The main mechanisms proposed include thermally driven circulation or the thermal bar (Bennett, 1971; Huang, 1971), the Stokes drift or the rectification of internal waves (Wunsch, 1973; Ou and Bennett, 1979), and the wind stress curl or the resultant heat redistribution from coastal upwelling (Emery and Csanady, 1973; Csanady, 1977).

In Lake Biwa, a wind stress curl has long been considered to be the main force maintaining the gyre system. For example, Suda *et al.* (1926) inferred that the prevailing wind was crucial to the formation of the gyre, and Hidaka (1927) carried out laboratory experiments to confirm this mechanism using a non-rotating barotropic model. Recently, Pu *et al.* (1987) analyzed the divergence and vorticity of the wind over and around Lake Biwa using hourly wind records from AMeDAS (Automated Meteorological Data Acquisition System), and found that a positive wind stress curl dominates the northern part of Lake Biwa throughout most of the year.

Unequal solar heat capture (mentioned above as topographic heat accumulation leading to thermally driven currents) is considered another important force (Oonishi, 1975; Ookubo *et al.*, 1984). Endoh (1986) evaluated the energetics of the First Gyre with a diagnostic model and found the maximum energy content occurred in September. From both, the theory and experiments on thermally driven currents, it can be deduced that the maximum energy content should occur in July when the buoyancy flux is at its maximum. Thus, the gyral currents in Lake Biwa develop mainly by thermal effects in May and are maintained by the effects of both heating and wind stress curl thereafter. In autumn, when the heating effect declines, the gyral currents remain energized by the persisting wind stress curl.

The First Gyre in Lake Biwa is remarkably stable and regular, very close to a true circle when compared with the cyclonic circulation of other large lakes. It is not yet clear why this is so, but a possible explanation lies in the horizontal scale and depth of Lake Biwa. Csanady (1978) pointed out that maximum current energy is present in an area contained by the internal radius of deformation or about 5 km in Lake Biwa. This is much smaller than the lake dimensions, thus the lake boundaries are far removed from the gyre.

The currents of the gyres play an important role in the transportation and mixing of dissolved or suspended materials, i.e. the metabolism of the lake. It is well known that the electrical conductivity of the lake water is almost uniform in the offshore zone of the North Basin of Lake Biwa, due to the intense mixing and dispersion of the water by the gyral currents. Recently, very fine bottom sediment was found in the area corresponding to the center of the gyres (Kamitani, 1988). The mechanism for transporting the mud into the gyre has not yet been well explained, but the large vertical circulation associated with the gyre (Endoh, 1986) no doubt plays an important role in the transportation and selection of suspended materials.

In summary, the gyres in Lake Biwa have the following properties, based on the above observations:

1. The First Gyre exists almost continuously from May to November: i.e. during the stratification period. In winter the gyres are not observed.
2. Gyre currents develop rapidly, in parallel with the thermal stratification, in May. The maximum energy content of the First Gyre occurs in September. Thereafter, the energy of the gyre decreases gradually as winter approaches.
3. Gyre is found only in the epilimnion.
4. The time-averaged current is in geostrophic balance including the centrifugal force, but the instantaneous current is greatly complicated by the effects of internal waves, inertial oscillation, and wind-driven currents.
5. The current velocity is approximately 0.1 $ms^{-1}$, however, much stronger currents (> 0.3 $ms^{-1}$ ) are sometimes observed in summer.
6. The shape of the First Gyre is very close to a circle, and the maximum current speed is observed at 3 - 4 km from the gyre center.
7. The locations of the gyres differ significantly from those estimated previously from non-averaged data.

# References

Bennett, J. R., (1971) Thermally driven lake currents during spring and fall transition period. *Proc. Conf. Great Lakes Res.* 14: 535-544.

Bennett, J. R., (1975) Another explanation of the cyclonic circulation of large lakes. *Limnol. Oceanogr.* 20: 108-110.

Csanady, G. T., (1977) On the cyclonic mean circulation of large lakes. *Proc. Natl. Acad. Sci. USA*, 74: 2204-2208.

Csanady, G. T. (1978) Water circulation and dispersal mechanism. In *Lakes - Chemistry, Geology, Physics*, edited by A. Lerman, Springer-Verlag, New York : 21-64.

Emery, K. O. and G. T. Csanady, (1973) Surface circulation of lakes and nearly land-locked seas. *Proc. Natl. Acad. Sci. USA*, 70: 93-97.

Endoh, S., (1976) Study on the water circulation in Lake Biwa. M.S. thesis, Kyoto Univ., Japan. (J)

Endoh, S., (1978) Diagnostic analysis of water circulations in Lake Biwa. *J. Oceanogr. Soc. Japan*, 34: 250-260.

Endoh, S., (1986) Diagnostic study on the vertical circulation and the maintenance mechanisms of the cyclonic gyre in Lake Biwa. *J. Geophys. Res*. 91C1: 869-876.

Endoh, S., S. Imawaki and H. Kunishi, (1979) Studies on the internal waves by observing the time variations of water temperature in Lake Biwa. *Annu. Disaster Prev. Res. Inst.* Kyoto Univ. 22B(2) 601-609.(J)

Endoh, S., I. Okamoto, (1981) Study on the circular currents in Lake Biwa. *Verh. Internat. Verein. Limnol.* 21: 136-140.

Endoh, S., I. Okamoto and M. Nakai, (1981) Circular currents in the North Basin of Lake Biwa (I) - seasonal variation of circular currents deduced from water temperature distribution. *Jpn. J. Limnol*. 42: 144-153.(J)

Endoh, S., I. Okamoto, R. Kishida and T. Kimura, (1982) Continuous records of water currents in Lake Biwa (I) - time variation of currents off Hikone and off Ado River. *Water Temp. Res*. 26-3: 35-47.(J)

Endoh, S., I. Okamoto, Y. Okumura, T. Tamura, K. Takano, Y. Hamai, T. Kodani, Y. Hayami, H. Asada, H. Kawamura and K. Iwane, (1987) Measurements of lake current by using radar. *Mem. Fac. Educ., Shiga Univ*. 37: 27-38.(J)

Endoh, S. and Y. Okumura, (1989) Continuous current measurements in Lake Biwa (II) - time variations of lake currents in the northern basin. *Jpn. J. Limnol.* 50: 341-350.(J)

Hidaka, K. (1927) Model experiment on the lake currents in the surface layer of Lake Biwa. *Bull. Kobe Marine Obs*. 13: 1-4.(J)

Huang, J. C. K., (1971) The thermal current in Lake Michigan. *J. Phys. Oceanogr*. 1: 105-122.

Imasato, N., S. Kanari and H. Kunishi, (1975) Study on the currents in Lake Biwa (I) - barotropic circular currents induced by the uniform wind. *J. Oceanogr. Soc. Japan,* 31: 15-24.

Imawaki, S., S. Endoh and H. Kunishi, (1979) Study on the water circulation in Lake Biwa - separation of the steady current and internal waves. *Annu. Disaster Prev. Res. Inst.* Kyoto Univ. 22B(2) 591-599.(J)

Imberger, J. and P. F. Hamblin (1982) Dynamics of lakes, reservoirs, and cooling ponds. *Ann. Rev. Fluid Mech*. 14: 153-187.

Itakura, Y., S. Endoh, I. Okamoto and M. Kumagai, (1990) Infrared remote sensing for study of water current of Lake Biwa. *Jpn. J. Limnol.* 51: 26-28.

Kamitani, T., (1988) Grain size distribution of muds and its depositional process in Lake Biwa. *J. Res. Gr. Clas. Sed.* Japan, 5: 23-33.(J)

Kanari, S. ,(1974) On the study of numerical experiments of two-layer Lake Biwa. *Jpn. J. Limnol.* 35: 1-17.

Kanari, S., (1975) The long-period internal waves in Lake Biwa. *Limnol. Oceanogr*. 20: 544-553.

Kunishi, H., I. Okamoto, and H. Sato, (1967) Observation of water circulation in Lake Biwa. *Annu. Disaster. Prev. Res. Inst.* Kyoto Univ. 10B: 321-329.(J)

Morikawa, M. and I. Okamoto, (1960) The surface currents of Lake Biwa deduced from drift-bottle experiments. *Jpn. J. Limnol.* 21: 173-186.(J)

Morikawa, M. and I. Okamoto, (1960) On the currents of Lake Biwa. *Mem. Fac. Educ. Shiga Univ*. 12: 21-30.(J)

Obayashi, S., S. Onishi, T. Maruyasu and Y. Takahashi, (1982) Study on application of remote sensing to limnological survey in Lake Biwa. *J. Remote Sensing Soc. Japan,* 2-2: 79-108.(J)

Okamoto, I., (1962) Some examples of the fluctuation of the water circulation in Lake Biwa. *Mem. Fac. Educ., Shiga Univ.* 12: 31-37.(J)

Okamoto, I., (1968) Time variation of water temperature in Lake Biwa (II) - variation of the temperature distribution accompanied with the fluctuation of water current. *Mem. Fac. Educ., Shiga Univ.* 18: 53-64.(J)

Okamoto, I., (1984) *Water currents in Lake Biwa,* edited by S. Horie, Dr. Jung Publishers, Netherlands: 175-183.

Okamoto, I. and M. Morikawa, (1961a) Water circulation in Lake Biwa as deduced from the distribution of water density. *Jpn. J. Limnol.* 22: 193-200.(J)

Okamoto, I. and M. Morikawa, (1961b) The determination of water circulation in Lake Biwa by the method of geopotential height. *Mem. Fac. Educ., Shiga Univ.* 11: 27-33.(J)

Okuda, S. and K.Yokoyama, (1975) Remote sensing technique applied to investigation of Lake Biwa. *Japanese Committee of Environmental Remote Sensing, Project Report*: 109-114.(J)

Okumura, Y., (1977) Recording current meter for the lake current measurements. *Bull. Osaka Electro-Communication Univ.* 13: 49-59.(J)

Okumura, Y. and A. Yamamoto, (1978) Observation of lake currents by the radio-tracking system of drifting buoys (II). *Bull. Osaka Electro-Communication Univ.* 14: 97-108.(J)

Okumura, Y. and S. Endoh, (1985) Continuous current measurements in Lake Biwa (I) -method and some results. *Jpn. J. Limnol.* 46: 135-142.(J)

Ookubo, K., Y. Muramoto, Y. Oonishi and M. Kumagai, (1984) Laboratory experiments on thermally induced currents in Lake Biwa. *Bull. Disaster Prev. Res. Inst., Kyoto Univ.* 34: 19-54.

Oonishi, Y., (1975) Development of the current induced by the topographic heat accumulation (I) - the case of the axis-symmetric basin. *J. Oceanogr. Soc. Japan*, 31: 243-254.

Oonishi, Y. and N. Imasato, (1975) Study on the currents in Lake Biwa (II) - barotropic responses to the uniform wind of a finite duration. *J. Oceanogr. Soc. Japan,* 31: 53-60.

Ou, H.W. and J. R. Bennett, (1979) A theory of the mean flow driven by long internal waves in a rotating basin, with application to Lake Kinneret. *J. Phys. Oceanogr.* 9: 1112-1125.

Pu, P., I. Okamoto, S. Endoh and M. Kumagai (1987) A study on the local wind system over lakes and the reasons of formation of lake circulation in case of Lake Biwa. Abstracts of *Int. Symp. on Lake Tai and Lake Biwa, Nanjing*: 211-216.

Suda, K., K. Seki, J. Ishii, S. Takaishi and S. Mizuuchi, (1926) The report of limnological observation in Lake Biwa (I). *Bull. Kobe Marine Obs. 8*, 104 pp.(J)

Wunsch, C., (1973) On mean drift in large lakes. *Limnol. Oceanogr.* 18:793-795.

# 3

# Field Observation in the North Basin

S. Endoh, Y. Okumura and I. Okamoto

## Introduction

As mentioned in the previous section, a great number of observations have been carried out in the last 30 years to clarify the structure of the currents associated with the gyres in the North Basin of Lake Biwa. In this section, some results from current measurements of the gyres are described. The two main observational techniques have been buoy-tracking and continuous current metering. These methods have steadily improved over the years and it is now possible to obtain accurate and dense data on the current field. Various materials have been used for drogues such as drift bottles, cross boards, drift cylinders, and window-shades. Buoy-tracking has been carried out using distance meters, magnetic compass, sextants, radio transceivers, radar, balloons, aircraft and satellites. By comparison, fixed current meters have a short history (about ten years) in Lake Biwa, but a great deal of information has already been accumulated using improved instrumentation and mooring methods.

## Drifter Tracking

In August of 1979, five buoys with window-shade drogues at a depth of 2 m were released in the central area of the North Basin around observation stations A-D, M, O, S. (Figure 3.1) and tracked by boat for several hours. A sextant was used to determine the positions of the buoys to an accuracy of about 100 m. The wind was rather weak

Physical Processes in a Large Lake: Lake Biwa, Japan
Coastal and Estuarine Studies, Volume 48, Pages 15–29

throughout the observation period. Figures 3.2a and 3.2b show the trajectories of thesebuoys on August 24 and 25, 1979 which indicate the existence of a large cyclonic gyre, the First Gyre. Current velocities ranged from 0.1 to 0.2 m $s^{-1}$ throughout the three day observation period. On August 26, eight buoys were tracked using two boats. Once again a stable cyclonic gyre was observed (Figure 3.2c).

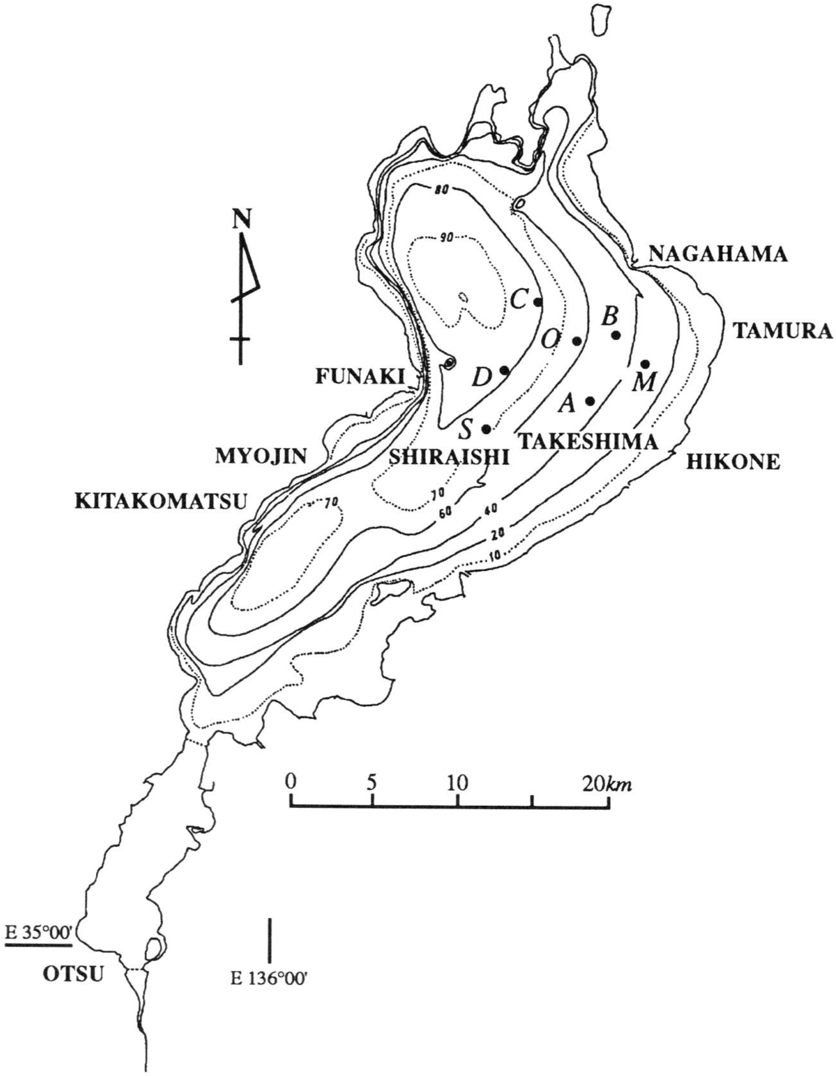


Figure 3.1. Map of Lake Biwa showing some observation stations. Depth contours are in meters.

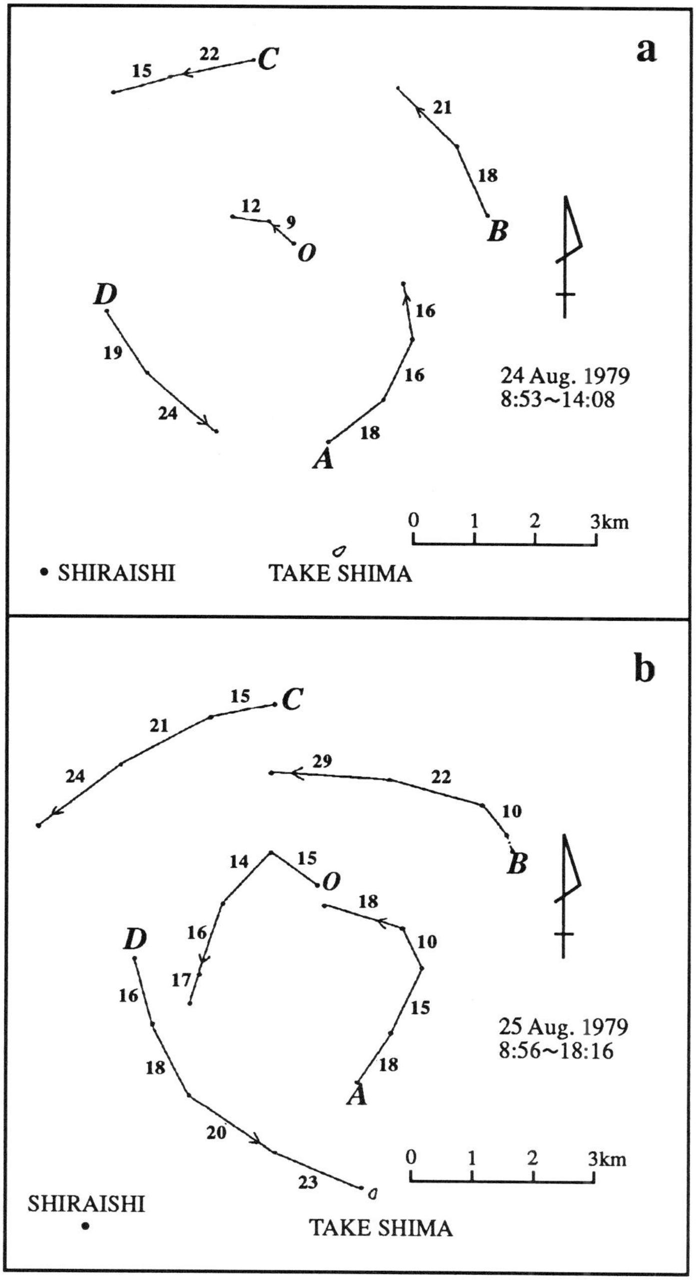


Figure 3.2. Trajectories of the buoys tracked by a boat with a sextant on August 24-26, 1979. The numerals beside the trajectories represent the current speed in cm $s^{-1}$.

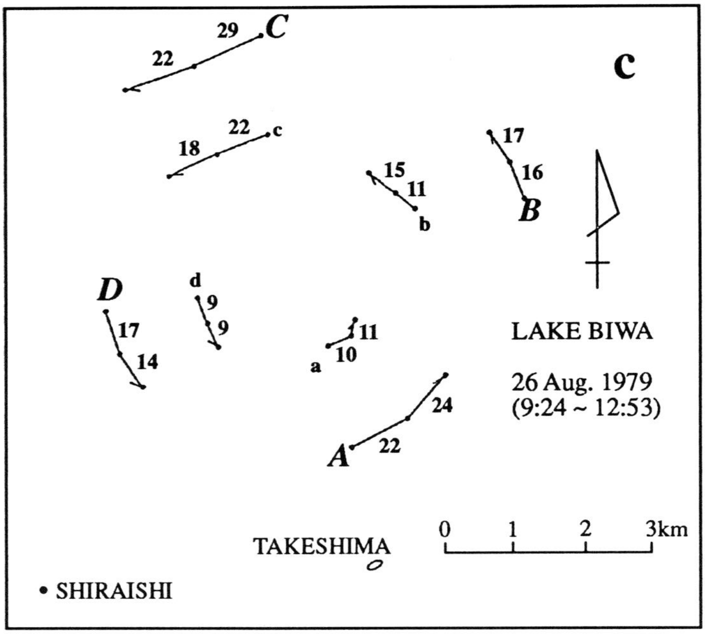


Figure 3.2 (Continued)

During the current observations, the three dimensional water temperature distribution was measured with a bathythermograph from another boat. Figure 3.3a shows vertical sections of water temperature along a cross section through Stations D, O and B, averaged over the period from August 24 to 25. A domed isotherm structure was observed within the gyre leading to a cold central core, corresponding to a geostrophic balance. The geostrophic current distribution calculated from this temperature distribution is shown in Figure 3.3b. The reference level was assumed to be located at the 30 dbar layer. Good agreement with the measurements shown in Figure 3.2 is apparent.

Figure 3.4 shows the dynamic height topography referenced to the 30 dbar layer and calculated using the water temperature distribution. The trajectories of the drifters are included to show that the drifter paths closely followed the dynamic height contours.

The horizontal divergence was estimated by calculating the time-change of the triangular area separated by three drifters. The result shows that the water in this gyre had a tendency to converge at a rate of $10^{-5}$ $s^{-1}$ similar to that estimated with the diagnostic model (Endoh, 1978). Figure 3.5 shows the vertical circulations calculated with a diagnostic model using the water temperature distribution shown in Figure 3.3a; two-cell circulations can be seen sinking near the gyre centre and spreading at the thermocline. Endoh (1986) pointed out that the resulting vertical circulation represented a decaying gyre. In general the vertical circulation will play an important role in metabolic and sedimentation processes and thus the distribution of suspended materials.

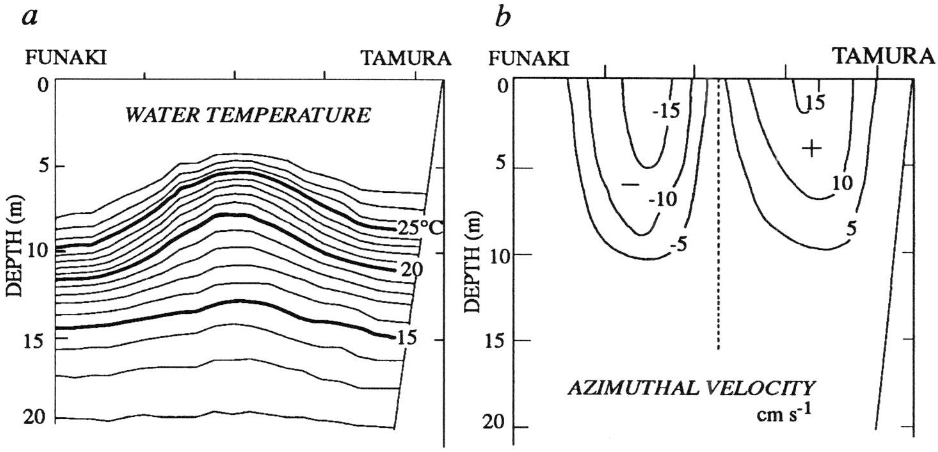


Figure 3.3. A vertical section of (a) water temperature averaged over August 24 and 25, 1979; (b) geostrophic current velocity calculated using water temperature with the reference level of 30 dbar.

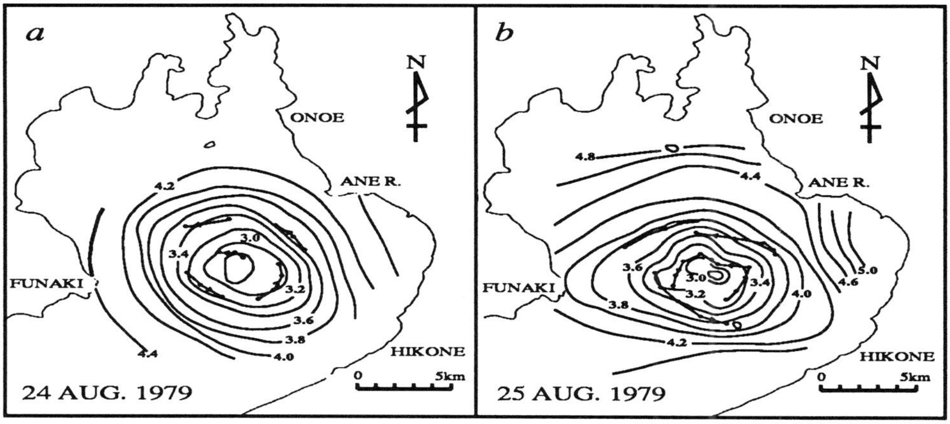


Figure 3.4. Dynamic height topography at the surface referenced to 30 dbar surface on (a) August 24 and (b) August 25, 1979. The trajectories of buoys are also shown.

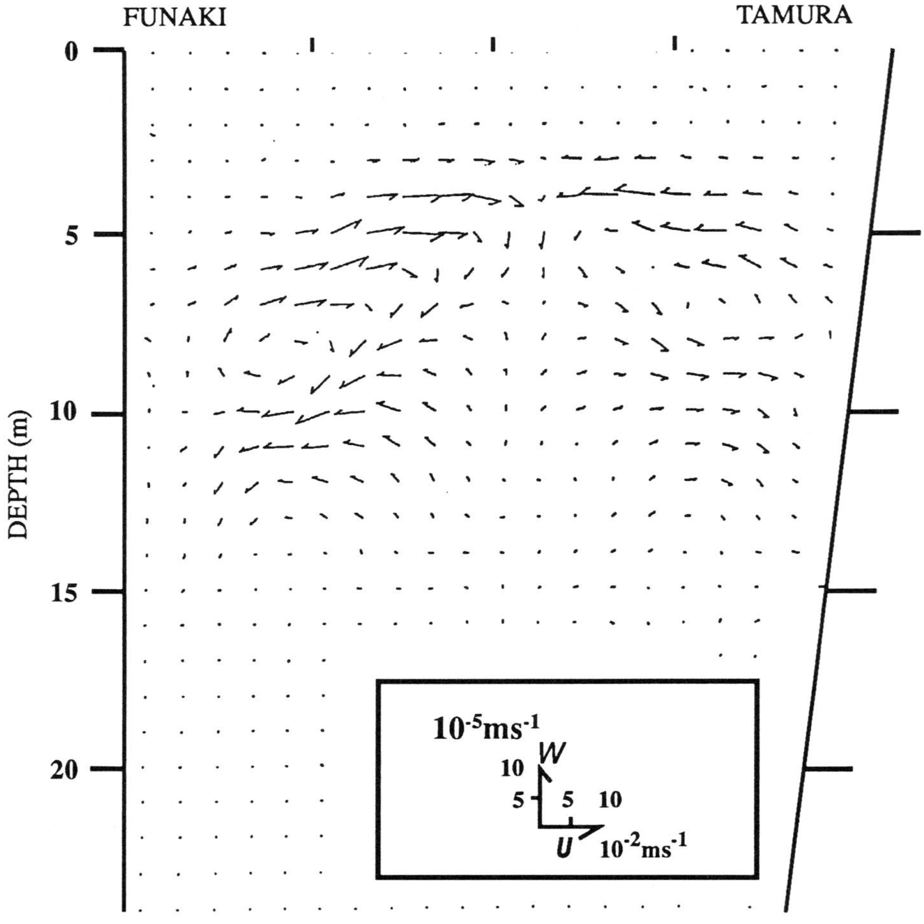


Figure 3.5. Vertical circulation in the gyre calculated by a diagnostic model.

Recently, drifter-tracking employing radar has been developed to obtain more continuous and accurate information on the current structure of the gyres (Endoh *et al.*, 1987). This method enables the simultaneous positioning of drifters with an accuracy of about 50 m even at night and in bad weather. The radar scanner was usually set on high ground beside the lake shore or on an island. Photographs of the radar display were taken at intervals of 30 minutes with an exposure time of about 10 seconds. A research vessel was employed to allow visual identification of the drifters. Each drifter was made from a window-shade which was connected by rope to a surface float, about 3 m long, with a corner reflector on the top. An example of a radar display is shown in Figure 3.6, taken at Takeshima Island in 1983. The several white points in the lake (the dark area) represent the echoes from the drifters.

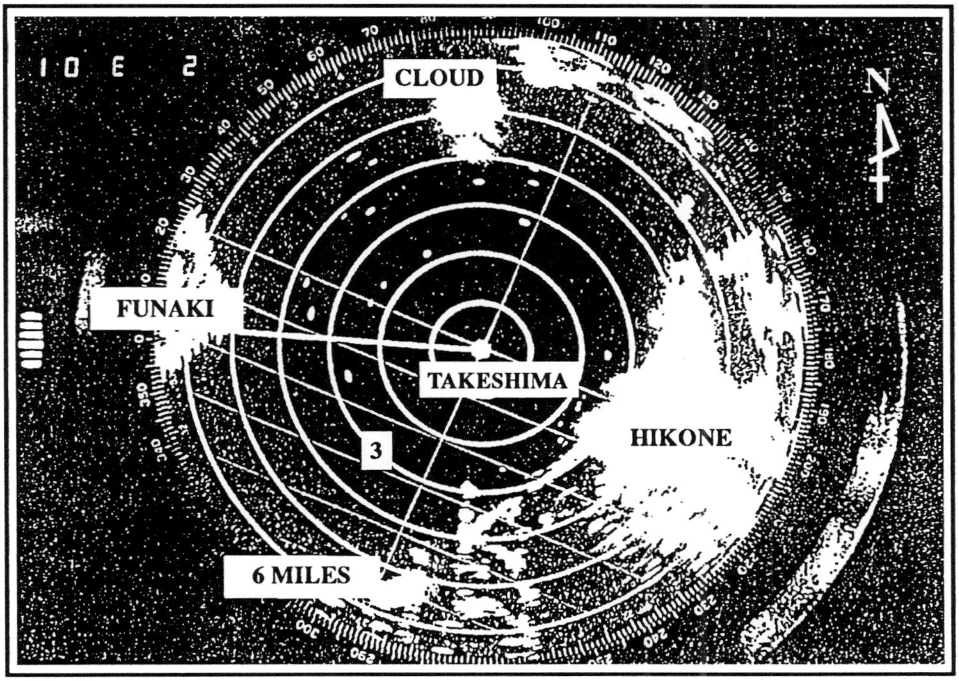


Figure 3.6. A picture of a radar display taken at Takeshima at 300 UT on September 2, 1983.

Figure 3.7 shows the results of drifter-trackings using radar in September, 1983 (after Endoh, 1986). The buoys which were drogued at a depth of 5 m (Figures 3.7a, 3.7b) rotated once every two days in a counter-clockwise direction, but some of them moved southward away from the gyre; the latter movements being caused by internal Kelvin waves which were also detected by the current meters (Endoh, 1986). Figure 3.7c shows four trajectories of drifters drogues at a depth of 10 m; these had tracks very close to a true circle due to the decreased effect of internal waves just above the thermocline.

Horizontal convergence was observed with every buoy at this depth, each moving towards the gyre centre with a radial velocity of about 5 x $10^{-3}$ m $s^{-1}$. Figure 3.7d shows the movements of deeper buoys (20 m and 30 m deep); here a very slow current speed was observed in the same direction as the surface layer. However, these deep drogues were influenced greatly by the overlying stronger currents. Once corrected for heavy drift (Endoh *et al.,* 1987), these drogues revealed that below the thermocline the water was almost stationary.

The Second Gyre (clockwise) has been frequently observed towards the south of the First Gyre. Because this gyre is anti-cyclonic (in a direction opposite to the earth's rotation), it is sometimes difficult to detect the structure of this gyre by water temperature distributions because of differences in the gyre dynamics. The pressure

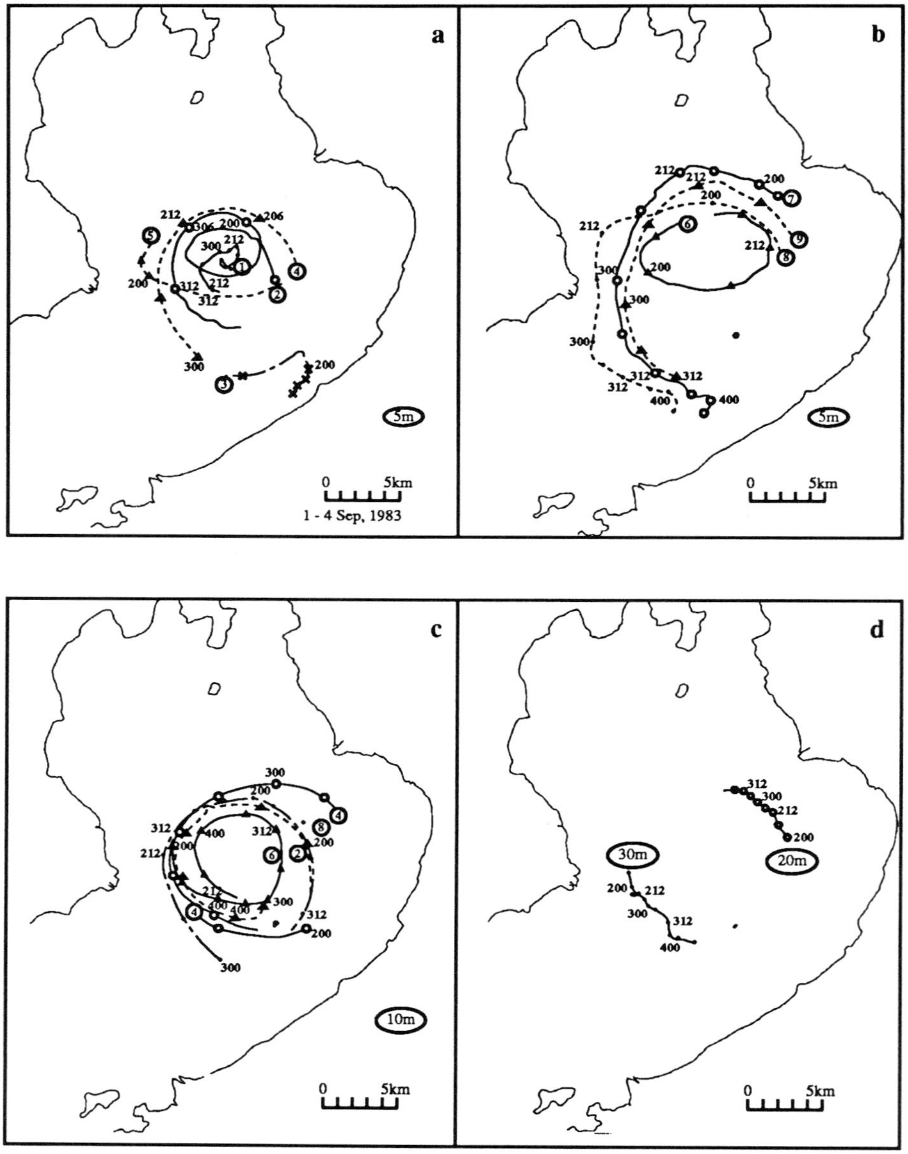


Figure 3.7. Trajectories of drifters at several depths on September 1-4, 1983. (a) and (b) at 5 m depth; (c) at 10 m depth; (d) at 20 m and 30 m depth. Numerals beside the trajectories show the day and hour e.g. 212 means 1200UT on September 2 (after Endoh, 1986; Endoh *et al.*, 1987).

gradient force (produced by temperature differences) is usually stronger in a cyclonic gyre because the Coriolis force and centrifugal forces are added. In an anti-cyclonic gyre the forces counteract each other resulting in a weaker feature.

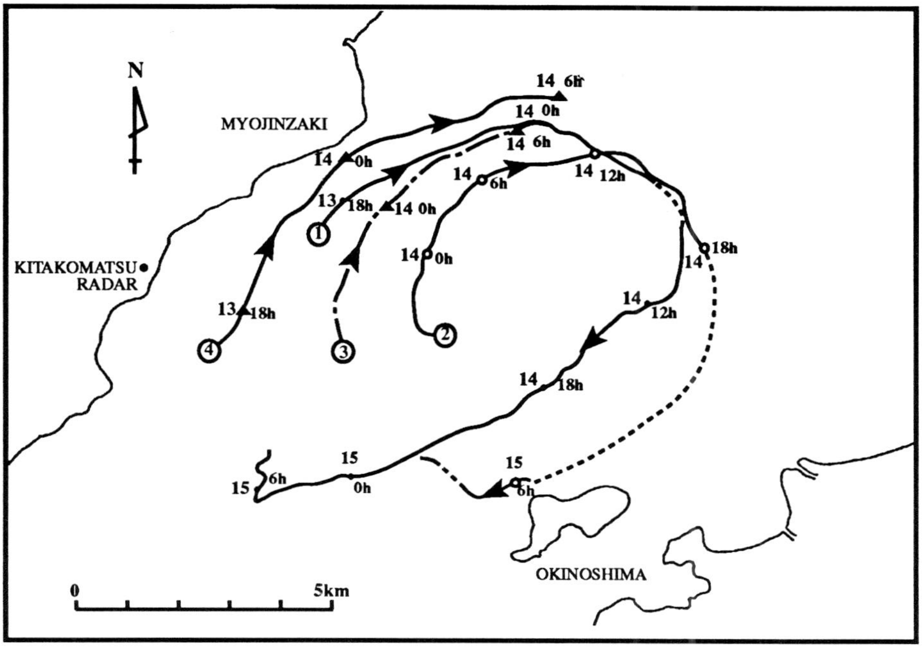


Figure 3.8. Trajectories of four drifters at a depth of 5 m showing the clockwise Second Gyre, September 13-15, 1986 (after Endoh *et al.*, 1987).

The best way to clarify the current structure of the Second Gyre is by direct current measurements. In September of 1986, several drifters were released and tracked by radar at Kitakomatsu. The method was almost the same as that described above. Figure 3.8 shows the trajectories of some of these drifters between September 13 and 15, 1986. All the drifters moved clockwise at a speed of between 0.1 and 0.2 m $s^{-1}$. The shape of this gyre was more elliptical, quite different from the circle-shaped First Gyre. The trajectories and current meters indicated the occasional development of a coastal jet, near Myojinzaki on the the western shore, with a maximum current speed of 0.4 m $s^{-1}$.

## Measurement by Current Meters

Since 1981, a number of current measurements have been carried out using automated current meters including the RCM-4 (Aanderaa) and DCM-3 (Osaka Electro-Communication University) (Okumura and Endoh, 1985: Endoh and Okumura, 1989). These current meters allow the continuous measurement of current speed, direction, and water temperature for about one month.

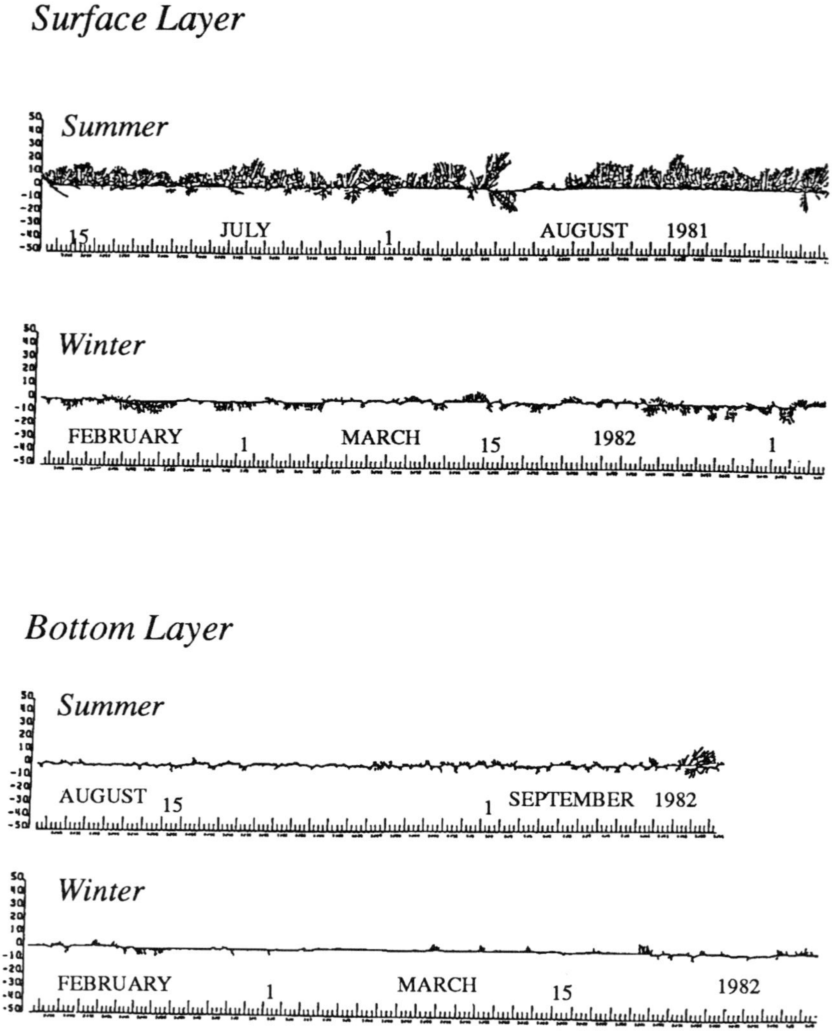


Figure 3.9 . Diagrams of time series of current vectors. Upper: the surface layer in summer and winter; lower: the bottom layer in summer and winter (after Okumura and Endoh, 1985).

The horizontal and vertical structure and time-variation of the current in Lake Biwa is somewhat unusual. In particular, the current field in winter and in the deep layer is poorly understood. Figure 3.9 shows the results of current measurements at the surface and at depth during summer and winter around Station M (Figure 3.1), about 5 km north-west of Hikone. Current vectors are shown, for each 10 minutes, for a period of about a month. In this figure, the strong and stable current in the surface layer in the summer is evident, as well as the relatively weak current in the winter. This is due to the existence

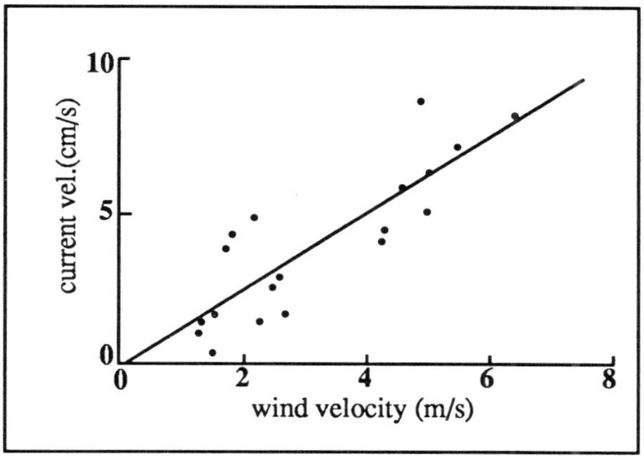


Figure 3.10. Scatter diagram of daily averaged current speed and wind speed. Correlation coefficient is 0.85 (after Okumura and Endoh, 1985).

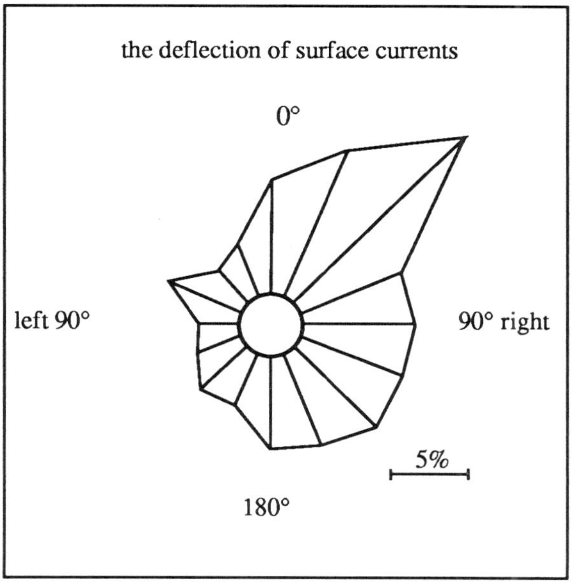


Figure 3.11. Frequency diagram of deflection of the current direction from the wind direction (after Okumura and Endoh, 1985).

of the gyre current in summer. In winter, no well defined gyre exists and the current field is therefore controlled by the wind. The correlation between the current and the wind velocity was high; the observed current was about 1/80 of the wind velocity (Figure 3.10). The surface current direction in winter was slightly deflected to the right of the wind direction by the formation of a weak Ekman drift (Figure 3.11).

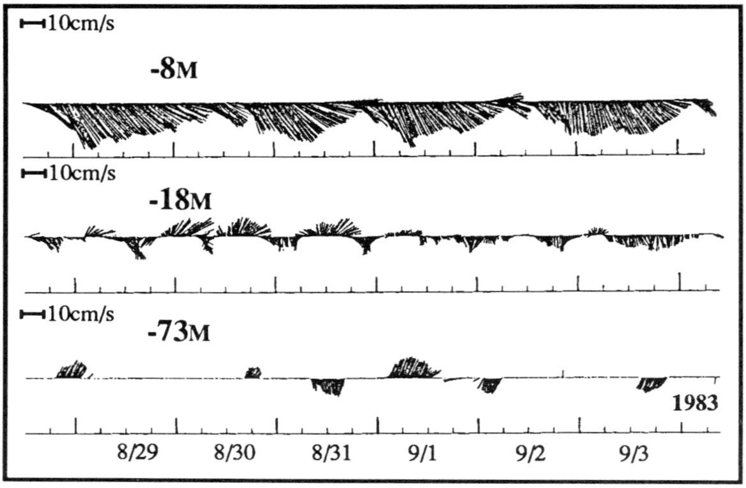


Figure 3.12. Stick diagrams of current in three layers (surface, thermocline and bottom) taken near Okino-shiraishi from August 28 to September 4, 1983 (after Endoh and Okumura, 1989).

In the deep layer, very weak currents were typically observed, although high velocities (0.2 m $s^{-1}$) sometimes appeared in the summer as a result of strong winds. Such deep strong currents are believed to be generated by vertical circulation or compensating currents caused by the winds or by internal waves. On the other hand, the deep layer in winter is quite calm even with wind conditions much stronger than those during summer.

Continuous current measurements have clarified the various oscillatory current structures including internal waves and inertial oscillation. Figure 3.12 shows the vertical current variation obtained at Station S near Shiraishi (Figure 3.1) during August 28 and September 4, 1983. The water depth was 74.5 m and the thermocline depth was about 15 m. The surface (8 m) current was strongly influenced by the First Gyre motion and had a long period of oscillation. The bottom (73 m) current was very weak and oscillated with the same period as, but out of phase with the surface current. The rotary current spectra are shown in Figure 3.13a, where it can be seen that the counter-clockwise rotating current was dominated by energy with a 43 hour period; indicative of the fundamental internal wave (internal Kelvin wave). On the other hand, the current variation at the thermocline depth (18 m) was very different from that in the surface and bottom layers. The period was shorter and the current direction varied in a clockwise fashion with time. The rotary spectra shown in Figure 3.13b indicate a steep spectral peak at 21 hour; the inertia period in the Lake Biwa basin. As shown in Figure 3.13a, there were no spectral peaks at this period in the surface and bottom layers. Following both observations it can be noted that the thermocline depth corresponds to the "level of no motion" for the long-period internal wave as well as marking the bottom of the gyre current. If a current happens to be created, the dominant force in this layer would seem to be unsteady inertia and Coriolis force, implying good conditions for the maintenance of inertial oscillations.

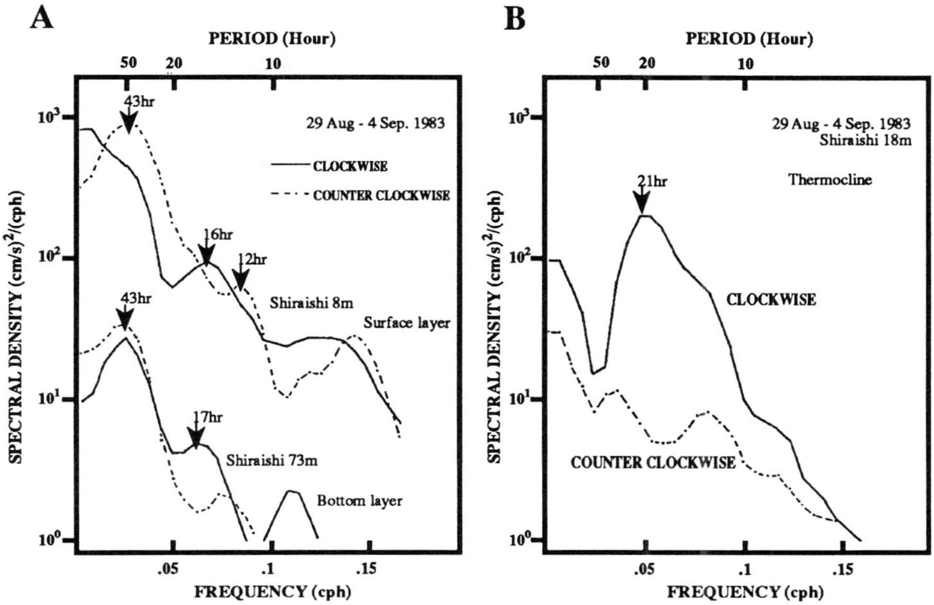


Figure 3.13. Rotary spectra of currents shown in Figure 3.12 (after Endoh and Okumura, 1989).

In the stratification period, short-period baroclinic oscillations occur frequently. The period of these oscillations depend on the intensity of stratification and were 17 hours in May and 11 hours in August (Figure 3.14). Csanady (1973) showed that the frequency $\sigma$ of the internal Poincaré wave in a channel of width $b$ may be expressed as:

$$\sigma = f\,(1 + R^2\,\pi^2/b^2)^{1/2},$$

where $f$ is the Coriolis' parameter (8.4 x $10^{-5}$ $s^{-1}$ )and R is the internal radius of deformation. The internal radius of deformation calculated using the observed water temperatures was 2.4 km in May and 6.2 km in August. The corresponding periods of the internal Poincaré wave was therefore, 18.5 hour and 12.6 hour respectively, almost coinciding with the observed periods.

# Conclusion

According to the recent current measurements in the North Basin of Lake Biwa, the cyclonic gyre (the First Gyre) is quite stable in the surface layer during the period of thermal stratification. This gyre sustains a weak large scale vertical circulation with sinking at the cool central core. This subsidiary circulation is considered to be a decay process in the gyre and plays an important role in the distribution of suspended substances in the lake. The Second Gyre to the south of the First Gyre is also stable but is characterized by an occasional coastal jet near the western shore. These exist only in the epilimnion and their strength is strongly correlated to the degree of stratification. Figure 3.15 shows the distribution of surface current vectors, averaged over about a month, that were obtained during mid-summer between 1981 and 1989; though the correlation has decreased, the typical summer circulation pattern can still be seen.

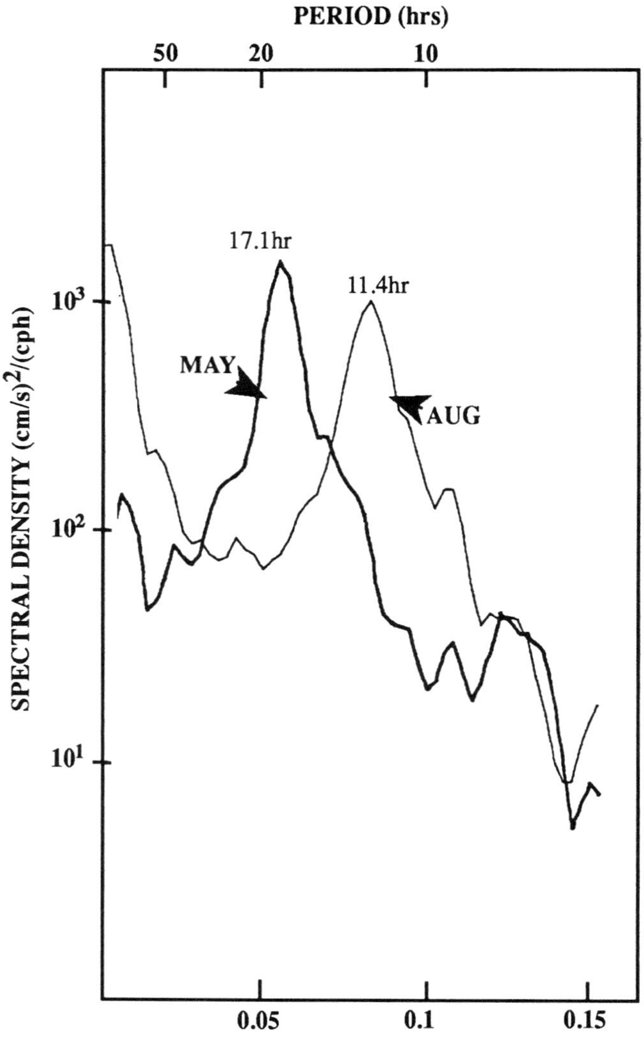


Figure 3.14. Rotary spectra (clockwise components) of the surface current in May and August. These spectral peaks correspond to the internal Poincare wave (after Endoh and Okumura, 1989).

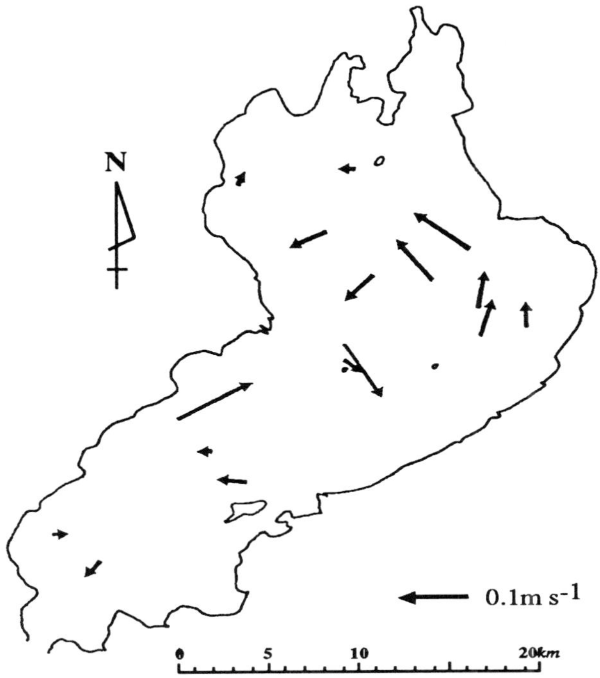


Figure 3.15. Horizontal distribution of time-averaged surface currents in the North Basin of Lake Biwa measured during the summer of 1981-1989.

# References

Csanady, G. T., (1973) Transverse internal seiches in large oblong lakes and marginal seas. *J. Phys. Oceanogr.*, 3: 439-447.

Endoh, S., (1978) Diagnostic analysis of water circulations in Lake Biwa. *J. Oceanogr. Soc. Japan*, 34: 250-260.

Endoh, S., (1986) Diagnostic study on the vertical circulation and the maintenance mechanisms of the cyclonic gyre in Lake Biwa. *J. Geophys. Res.*, 91C1: 869-876.

Endoh, S., I. Okamoto, Y. Okumura, T. Tamura, K. Takano, Y. Hamai, T. Kodani, Y. Hayami, H. Asada, H. Kawamura and K. Iwane, (1987) Measurements of lake current by using radar. *Mem. Fac. Educ., Shiga Univ*. 37: 27-38.(J)

Endoh, S. and Y. Okumura, (1989) Continuous current measurements in Lake Biwa (II) - time variations of lake currents in the northern basin. *Jpn. J. Limnol.* 50: 341-350.(J)

Okumura, Y. and S. Endoh, (1985) Continuous current measurements in Lake Biwa (I) - method and some results. *Jpn. J. Limnol*. 46: 135-142.(J)

# 4

# Water Mass Exchange Between the Main Basin and Shiozu Bay

I. Okamoto and S. Endoh

## Introduction

Shiozu Bay is the largest bay in Lake Biwa and is situated at the northern end of the main basin. The bay is 7 km long, 3 km wide, has a mean depth of 30 m, and its volume is 6.2 x $10^8 m^3$ (Figure 4.1). The bay is surrounded by beautiful scenery and historical sites, such as Chikubu Island (with an old temple and shrine), the Tsuzurao-saki archeological site (6th cent. B.C.), the old battle site of Mt. Shizugatake, and the old Nishino Tunnel (1845). A limnological survey of the bay began in 1968, when it was selected as one of the representative oligotrophic lakes in Japan by the National Committee of IBP (International Biological Program), and the bay subsequently served as a field site for training students doing limnological observations.

The bottom topography of the bay is characterized by a steep slope on the western shore, a gradual slope on the eastern shore and a well-developed delta at the bay mouth near Katayama and Onoe, as shown in Figure 4.1. An internal seiche or standing free oscillation of the thermocline can often be observed in the bay (Okamoto and Iwamato, 1977). The oscillation period depends on the seasonal thermal structure. In the summer, the oscillation period coincides with the period of the daily wind cycle, so that an internal seiche often develops in summer under resonant forcing. A second internal oscillation with a 43-hr period ( in summer), originally formed in the main basin and associated with Kelvin waves, is also observed in the bay. Both internal oscillations

Physical Processes in a Large Lake: Lake Biwa, Japan
Coastal and Estuarine Studies, Volume 48, Pages 31–42

have a significant influence on the water movement in the bay and make a considerable contribution to the water mass exchange between the main basin and the bay (Okamoto, 1985). A surface seiche is also evident in the bay and contributes similarly to the exchange flow but to a smaller extent. Another mechanism for mass exchange is a vertical circulation caused by the prevailing winds.

## Observations

Exchange of water mass was studied by evaluating the changes in temperature and current velocity in the bay at measurement sites shown in Figure 4.1. Two methods of temperature observation were usually employed concurrently: 1) long-term continuous measurement by automatic recording instruments at anchored stations and 2) repeated sounding of the vertical temperature distribution along the longitudinal section extending from the bay head to Chikubu-Island. These temperature observations were used to indirectly estimate the water mass exchange.

Current velocity observations were also made by two techniques: an Eulerian method comprising long-term continuous measurement of current velocity at anchored stations, and a Lagrangian method involving the tracking of drifting drogues with the aid of theodolite fixes and a radarscope.

Lake level gauges were installed at the bay head (Hannoura) and near the bay mouth (Katayama) providing valuable information on the water mass budgets and lake surface oscillation due to surface seiches. The wind velocity was continuously measured at Onoe Port and opportunistically on boats.

## Seiches and internal seiches

An example of drift trajectories is shown in Figure 4.2. A drogue in the 1 m-layer drifted southwards, while the others (20 m- and 53 m-layer) moved northwards. Since the thermocline depth was 15 m, the two deep drogues were below the thermocline and the trajectories were consistent with an internal seiche motion. However, the drift distances over a period of 5 minutes were not constant but changed from time to time. Drift velocities were calculated at intervals of 5 minutes and their components parallel to the basin axis are plotted against time in Figure 4.3 (positive: northwards, negative: southwards). The velocity components at the three depths exhibit a high frequency oscillation with a period of 37 minutes. This probably corresponds to the first mode of free oscillation for the bay. The lake surface elevations at Hannoura and Katayama are shown in the upper part of Figure 4.3. These elevations show the same oscillation period as those exhibited in the velocity components. We can also see from the phase differences that the velocity components were a quarter of a period (9 minutes) earlier than the surface elevations. This is consistent with a seiche motion. Another seiche at Katayama was caused by the shelf water oscillating in the delta near the bay mouth.

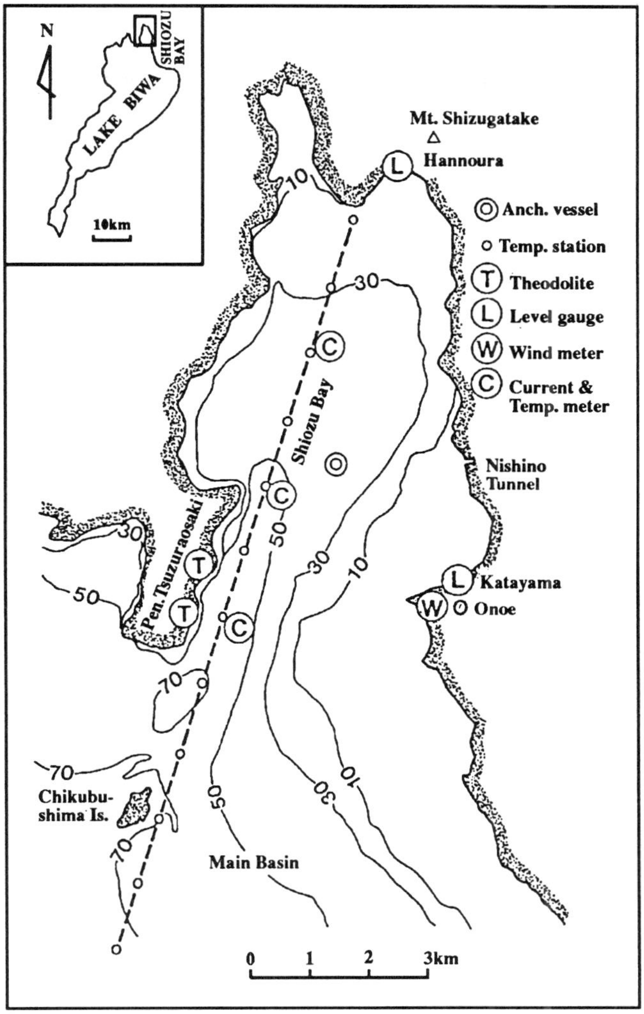


Figure 4.1. Location of observation stations in Shiozu Bay, Lake Biwa. Depth contours in meters.

Temperature oscillations collected near the bay head at the most shoreward temperature station are shown in Figure 4.4. Here the thermocline extended from a depth of 10 m to 20 m and both temperature sensors were located in the thermocline. The water mass budget across the bay mouth may be estimated from hypsographic data shown in Figure 4.5. Temperature oscillations at fixed depths are caused by periodic vertical movements of the thermocline. For instance, a temperature rise corresponds to a downward movement of the thermocline, while a temperature fall corresponds to an upward movement of the thermocline. Vertical displacement of the thermocline may be written as follows:

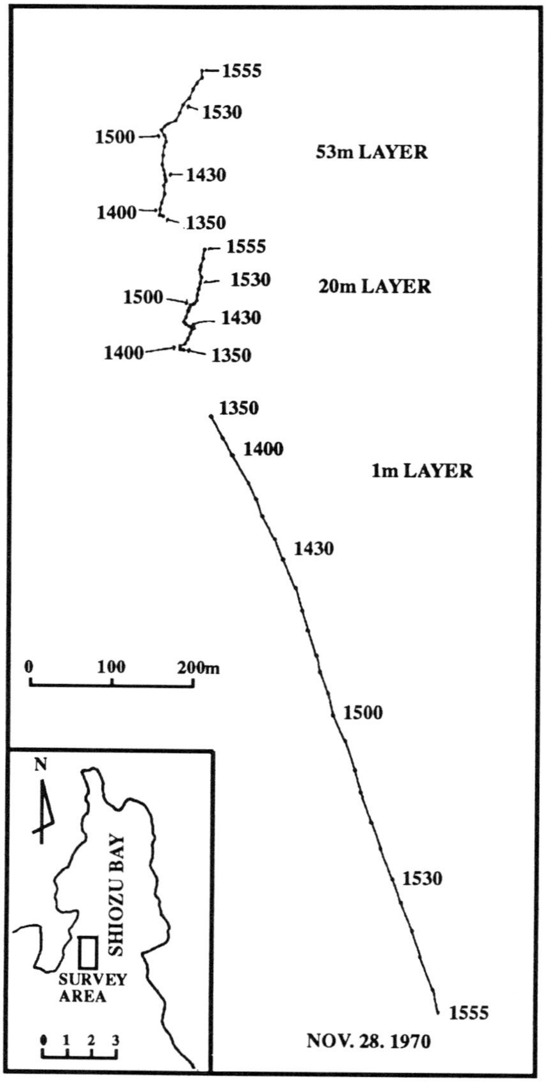


Figure 4.2. Drift trajectories tracked by theodolites at five minute-intervals. Numbers show the observation times.

$$\delta = \Delta T(t)\left(\frac{dT}{dz}\right)^{-1}$$

where $\Delta T(t)$ is the temperature variation with time, and $\left(\frac{dT}{dz}\right)$ is the vertical gradient of temperature.

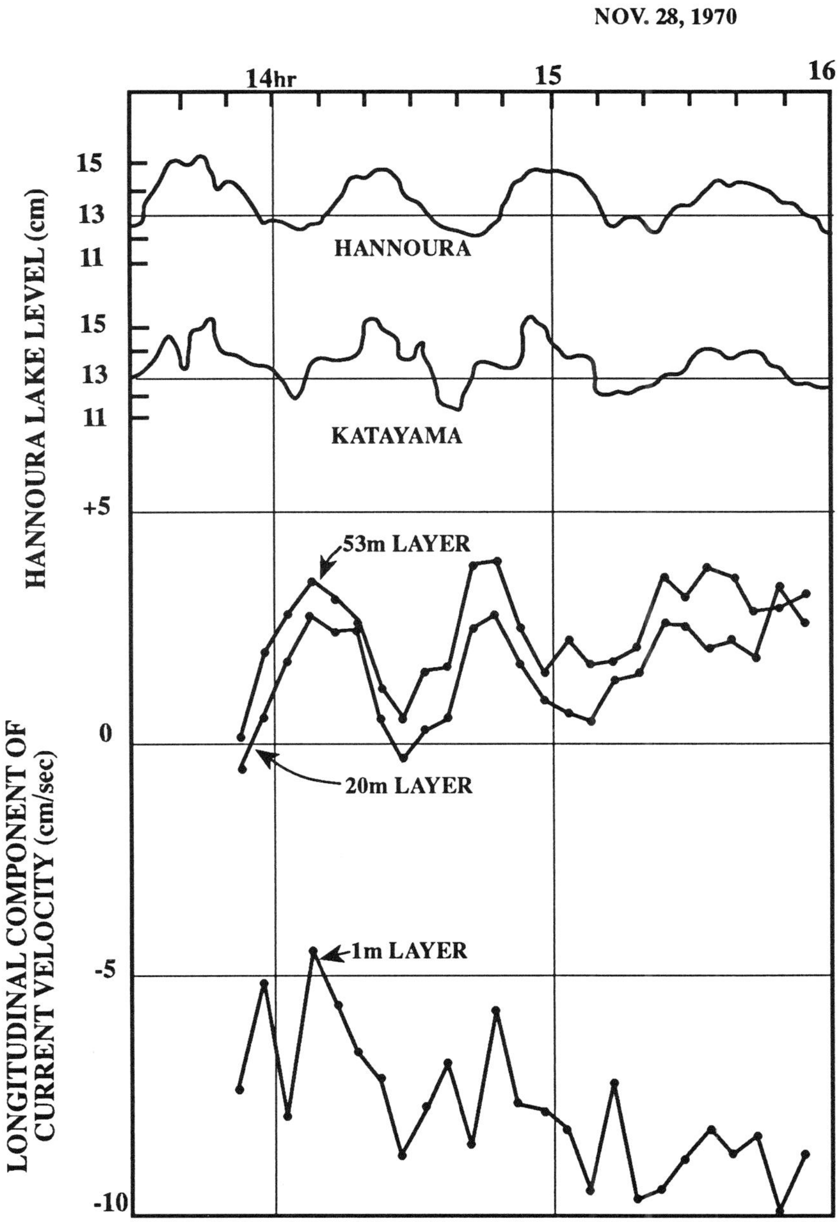


Figure 4.3. Longitudinal components of current velocities at intervals of five minutes calculated using Figure 4.2, positive: northward, negative: southward. The corresponding lake surface elevations at the two stations are shown in the upper section.

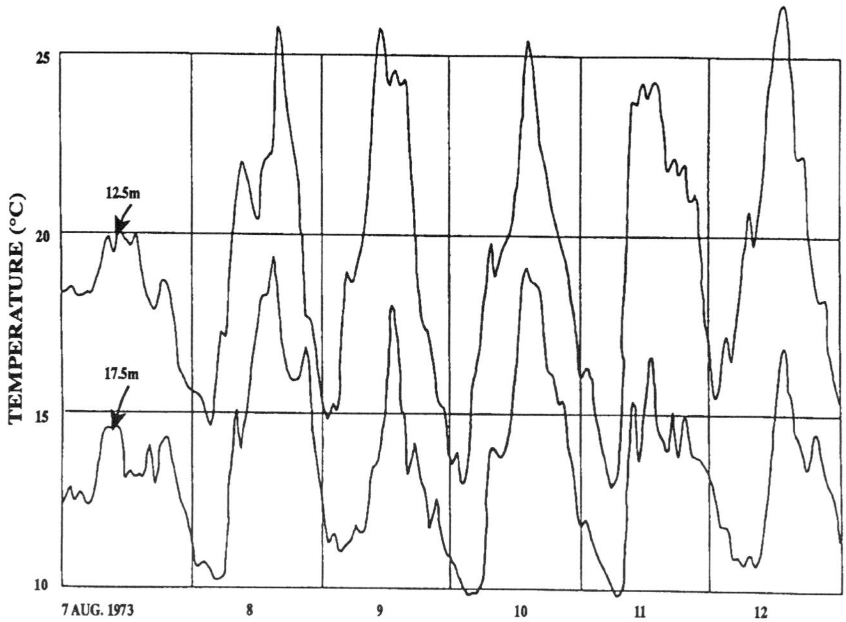


Figure 4.4. Example of periodic oscillation of temperature (internal seiche) with a 24-hr period observed at the bay head, at the most northern temperature station shown in Figure 4.1.

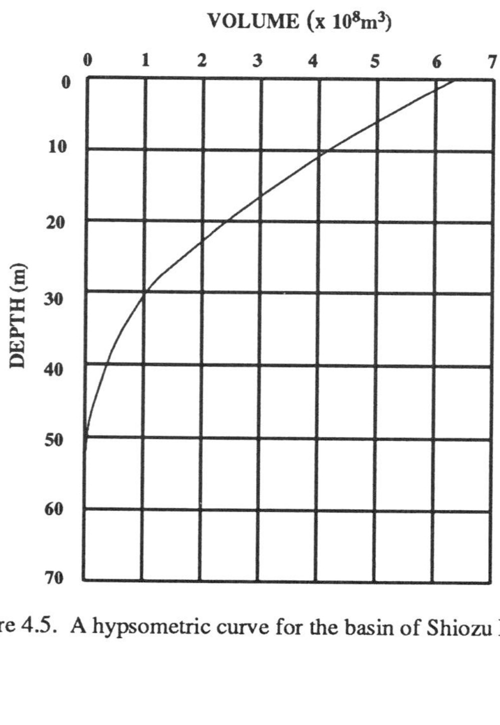


Figure 4.5. A hypsometric curve for the basin of Shiozu Bay.

Given the data in Figure 4.4 this leads to a vertical excursion $\delta$ of order 10 m.

The ratio of the internal seiche amplitude at the bay head to the bay mouth is equal to 1 : 0.4 in Shiozu Bay (Okamoto and Iwamoto, 1977). Therefore, the thermocline displacement was 4 m at the bay mouth or an average of 7 m over the entire bay basin, implying the average vertical displacement of the thermocline extended from a depth of 11.5 m to 18.5 m. We can estimate the water mass exchange caused by this vertical displacement of the thermocline by means of a hypsometric curve for the Shiozu Bay basin (Figure 4.5). We see the water mass volume corresponding to the depths of 11.5 m and 18.5 m is $4.0 \times 10^8 m^3$ and $2.6 \times 10^8 m^3$. As a consequence, the mass exchanged during half the period was estimated to be $1.4 \times 10^8 m^3$ which is equivalent to 22% of the total bay water mass. The water mass of $1.4 \times 10^8$ $m^3$, however, does not leave the bay basin but mostly returns, as internal seiches are periodic.

Figure 4.6 shows a time series of the temperature and longitudinal velocities at four anchor stations. The dominant temperature oscillation period was about 2 days corresponding to the period of the internal Kelvin wave that had developed in the main basin. Temperature oscillations were consistently in phase with each other while the current velocities at 8 m and 43 m were 180° out of phase with the temperature signal. As the upper layer water moved northward and the lower layer southward, the temperature at all stations rose simultaneously, and conversely, when the upper layer flowed southward and the lower layer northward, the temperature fell. Finally, at Station B, where the water depth is 45 m, the current velocity was as high as 20 $cms^{-1}$ in the lower layer (only 2 m above the bottom), suggesting that the bottom topography of the semi-closed bay mouth may intensify the seiche induced currents due to shoaling.

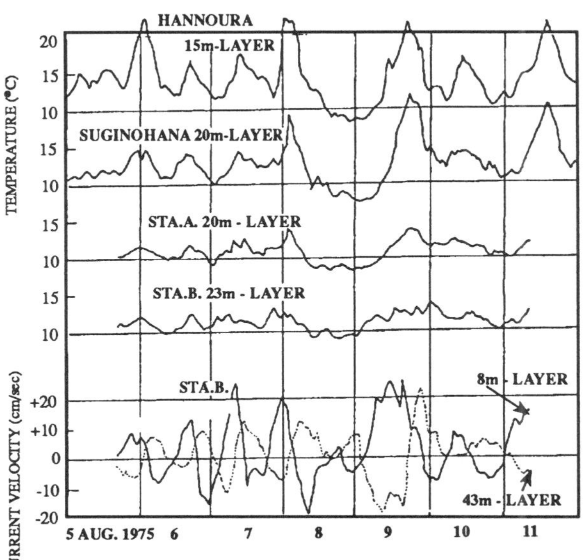


Figure 4.6. Periodic oscillations of temperature at four stations (upper) and of longitudinal components of current velocities in two layers (lower); northward current is positive.

Shiozu Bay has two internal seiches, one of which is a free oscillation of the bay water and the other is forced by the internal Kelvin wave intruding from the main basin. Figure 4.7 shows the result of a Fourier analysis of current velocities (a1-a3), wind velocity (b), and temperature (c1-c4). Dominant amplitude peaks can be seen in all three, with a common period of about 1 day (24-27 hours). A second peak occurred at a period of about 2 days (42-44 hours), and these two peaks correspond to the two internal seiches. In the 20 m-layer the current amplitude was exceptionally small at the 24 hour period (a2) because the 20 m-layer was just within the thermocline. The small amplitude in the 30 m-layer temperature was caused by a weak thermal stratification (c4), and the 24 hour period surface temperature was a consequence, not of the internal seiche, but rather of the diurnal heat budget.

An example of drift trajectories is presented in Figure 4.8. Four drogues were released into the surface layer (8 m depth) on the morning of August 8, 1986 and tracked by radar at intervals of one hour. Immediately after the release, all the drogues drifted toward the head of the bay, changed their directions at about 13:00, and then drifted south toward the bay mouth. After half a day, one drogue turned again at 4:00 the next morning and continued to drift back towards the head of the bay, while the others were caught in set fishing nets. The two direction changes are attributable to the 24 hour-period internal seiche. As the first turn occurred at 13:00 on August 8, the second turn was expected at 1:00 on August 9. However, it actually occurred 3 hours later, possibly due to the additional influence of the north wind at that time. As shown in Figure 4.9 the first turn occurred when the water temperature was at its highest and the second turn occurred almost at the time of the lowest temperature. Finally, it should be pointed out that immediately before and after the turn the recorded temperature continuously showed large and frequent fluctuations suggesting that large-scale mixing was taking place. The above scenario is very similar to that described by Monismith *et al.* (1990).

## Vertical circulation

When a persistent north wind blows, it is observed to push the upper layer of water out of the bay and set up a deep compensating flow from the main basin into the bay basin leading to upwelling. Due to the influence of an approaching typhoon, there was a strong north wind during the observation period from August 4 to 7, 1986. On August 6, 1986, when the observations began, the surface temperature obtained from an anchored vessel at the bay center (Figure 4.1) fell to a surprising low of 20 °C, as shown in Figure 4.9. This was 7-8 °C colder than usual, as normally the 20 °C isotherm surface lies 15 m below the lake surface in August; all the warm surface water had been removed from the bay basin. This volume amounted to about 2.8 x $10^8 m^3$ or 45% of the total bay water mass.

When the north wind ceased the warm water returned from the main basin into the bay basin, leading to downwelling of the upper layer water. Thus, water mass exchange occured again and the process of temperature recovery is shown in Figure 4.9.

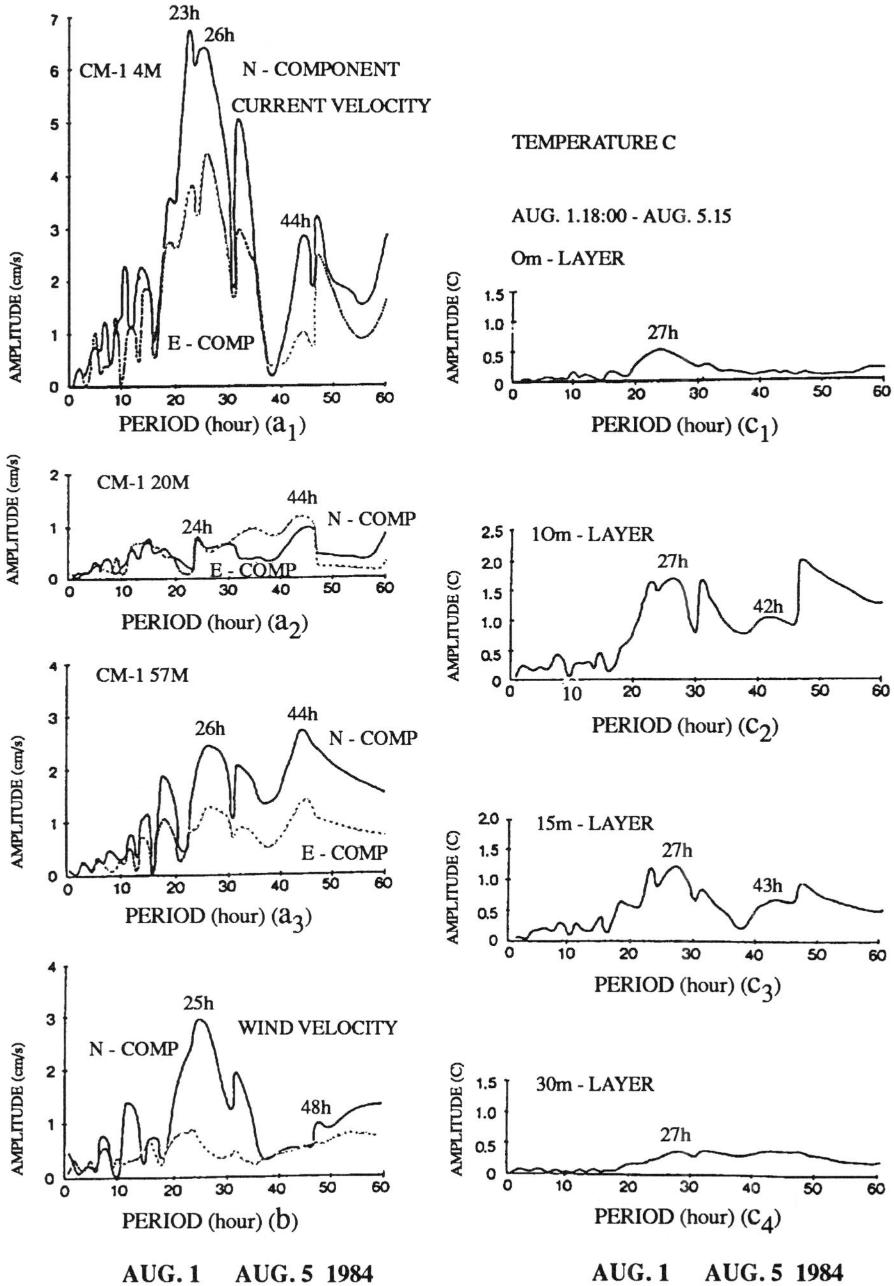


Figure 4.7. Amplitude spectra of current velocity components ($a_1$ - $a_3$), wind velocity components (b) and temperature ($c_1$ - $c_4$) in Shiozu Bay.

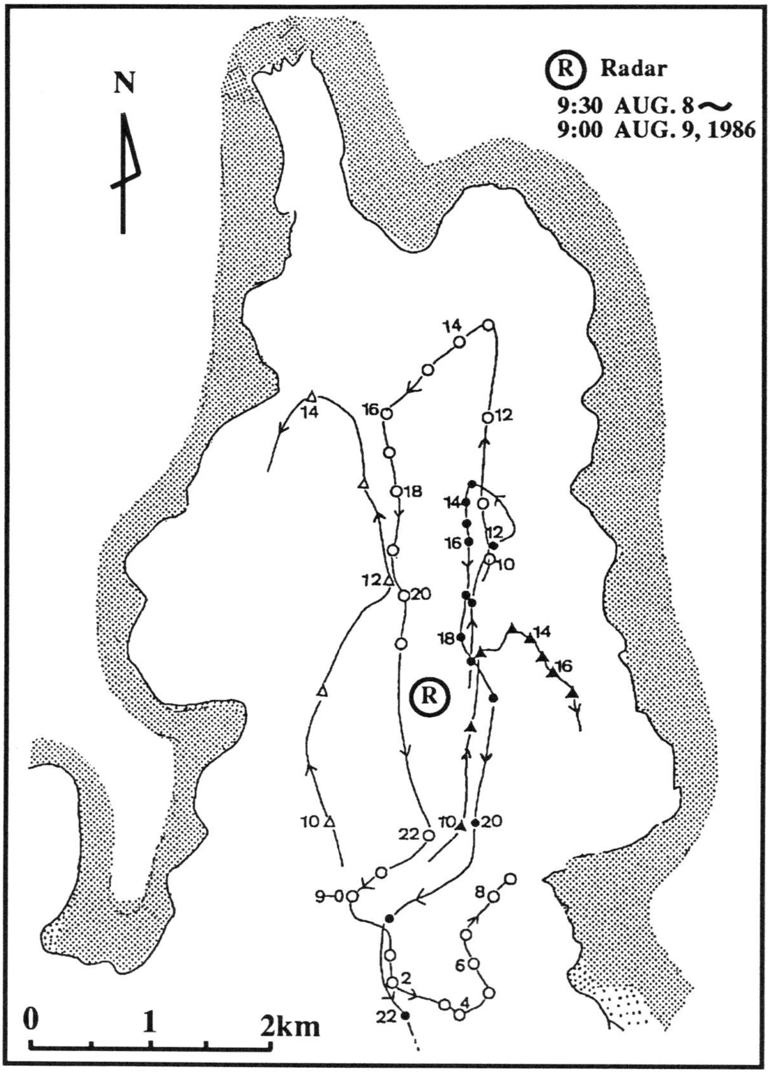


Figure 4.8. Drift trajectories tracked by a radar beacon at one-hour intervals. Numbers show the observation times in hours.

The upper layer temperature rose rapidly due to the return of warm water from the main basin and in the 20 m layer where the thermocline was located, a 1 day-period internal seiche occurred. Shown in Figure 4.10 are the current velocity distributions observed in a longitudinal section on the morning of August 8, 1986, at which time the temperature rose in all the layers as shown in Figure 4.9; the velocity distribution shown in Figure 4.10 was consistent with the events described above.

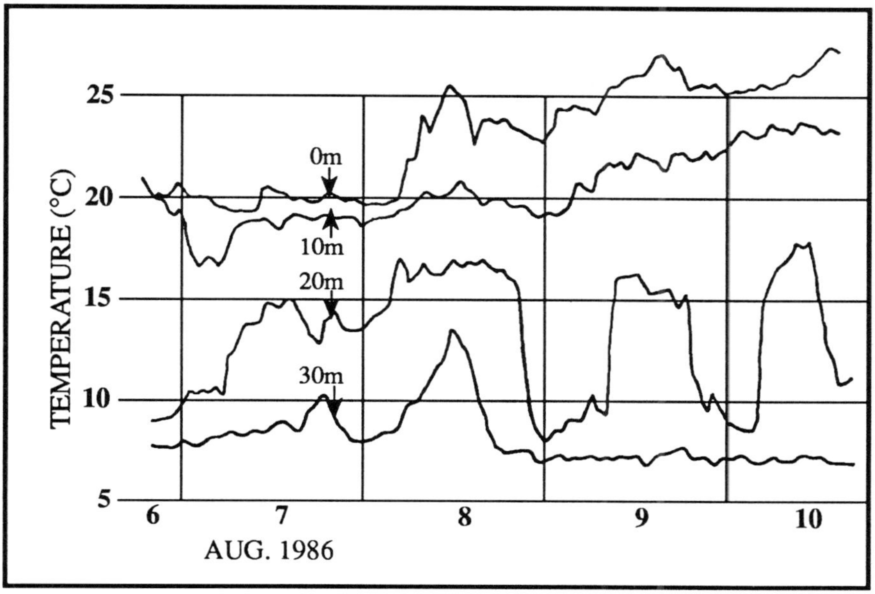


Figure 4.9. Processes for temperature recovery observed from an anchored vessel after cessation of the strong north wind.

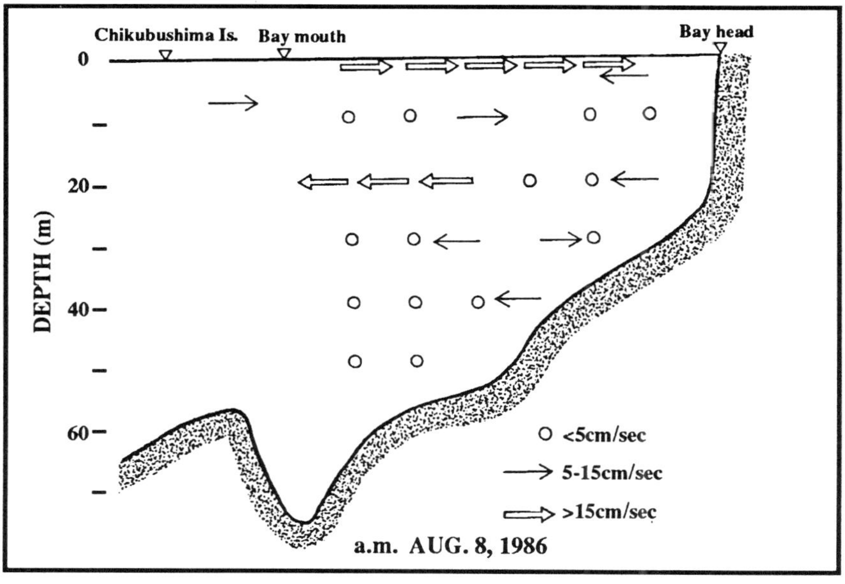


Figure 4.10. Distribution of longitudinal components of current velocities immediately after cessation of the strong north wind.

## Summary

Shiozu Bay is the largest bay in Lake Biwa situated at the northern end of the main basin. Water mass exchanges between the main basin and Shiozu Bay are largely attributed to two kinds of internal oscillations, to a surface seiche and to wind-driven vertical circulation. The first internal oscillation was a standing wave in the bay on the thermocline (internal seiche). The period was 24 hours in the summer season and was often developed due to resonance with a daily wind. The exchange volume was estimated by using well defined temperature oscillations recorded simultaneously at two different water depths. The volume was 1.4 x $10^8 m^3$ equivalent to 22% of all bay water per half period. The coefficient of horizontal diffusion driven by this oscillatory motion was estimated at 2.1 ~ 4.1 x $10^5 cm^2 s^{-1}$. The second internal oscillation occurred originally in a similar way as the first oscillation. Power spectra show the peaks in temperature and current velocity recorded simutaneously in summer at a period of about 1 day (1st oscillation) and about 2 days (2nd oscillation). A characteristic feature of the internal oscillation was that the flowing water mass in the upper layer, above the thermocline, was always compensated by the counter-flowing water mass of the lower layer. Strong northerly winds forced the warm surface layers from the bay causing an inflowing and upwelling of cold deep main basin water. This resulted in a water mass exchange of 45% of total bay water.

## References

Okamoto, I. and N. Iwamoto, (1977) Current observations in Shiozu Bay, Lake Biwa (V). *Mem.. Fac. Educ. Shiga Univ.* 27: 44-54. (in Japanese with English abstract)

Okamoto, I., (1985) Water movements as deduced from temperature distribution and its time variation in Lake Biwa. *Mem.. Fac. Educ. Shiga Univ.* 35: 19-34.

Monismith, S., J. Imberger, J. and M. L. Morison, (1990) Convective motions in the sidearm of a small reservoir. *J. Limnol. Oceanogr.*, 35(8): 1676-1702.

# 5

# Field Observations of the Dense Bottom Current Between the North and South Basins

K. Okubo

Water quality differs greatly in the North and South Basins of Lake Biwa, making the nature of water exchange between the basins a subject of considerable interest. In order to understand this exchange, monitoring of the water temperatures and flow velocities normal to the cross-section was performed at the lake's narrowest point (the dividing line between the North and South Basins), at the Lake Biwa Bridge (hereafter referred to as LBB; see Figure 5.1). Some results of this monitoring are described in the following section. It was observed that the shallower South Basin becomes cooler than the North Basin due to differential cooling, and northward-flowing dense bottom currents are observed at the LBB section in November and December. However, even during these months, the flow is intermittent with currents persisting only for a few days under calm conditions and attaining full strength 2 - 3 days after each major cooling event. Light surface currents could be expected in the summer, however, there is a constant southward advective current, due to river outflow from the southern end of of the South Basin on which the dense bottom current must be superimposed.

When the bottom current does occur, a temperature front is observed on the surface layer 2 - 3 km south of the LBB, as shown in Figure 5.2. Water from the North Basin flows into the South Basin in the upper layer along the western shore, while isotherms rise toward the eastern shore close to the LBB (Figure 5.3). The distortions of the temperature field are subtle, but are likely due to the effects of the earth's rotation. As will be shown in Chapter 6, numerical experiments by Oonishi *et al.* (1980) suggest that the flow can be characterized by the Rossby number,

Physical Processes in a Large Lake: Lake Biwa, Japan
Coastal and Estuarine Studies, Volume 48, Pages 43–51

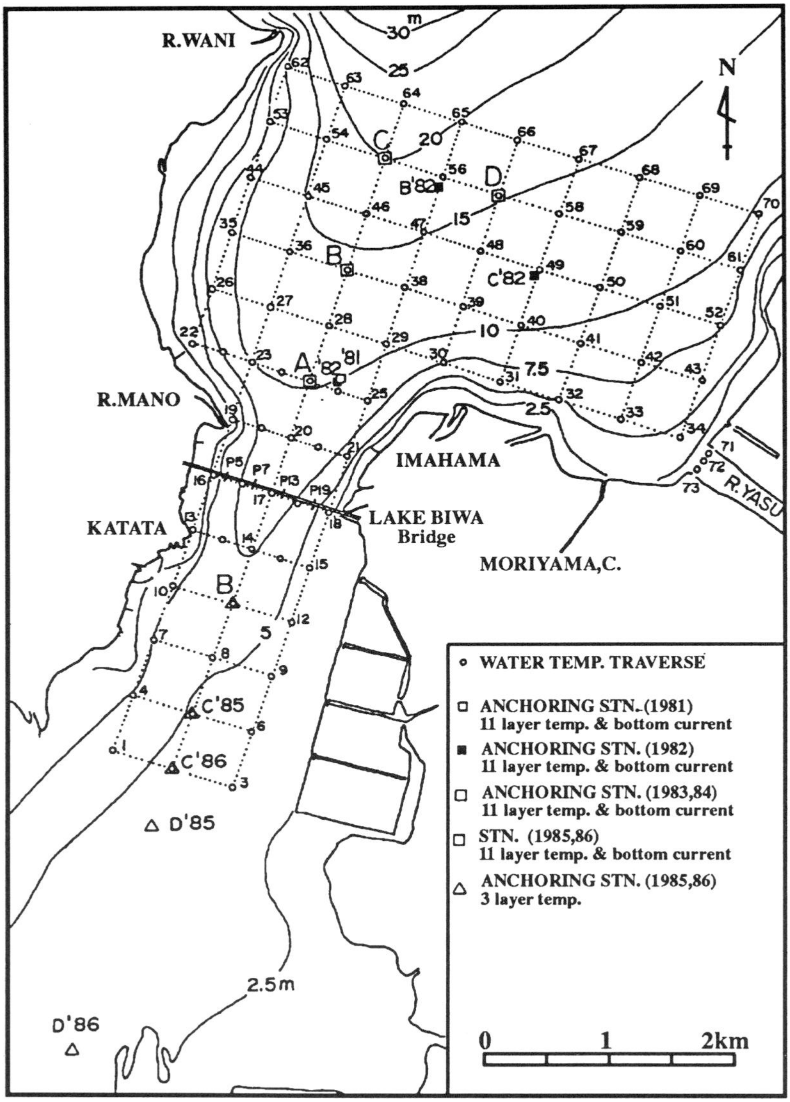


Figure 5.1. Map of the basin around Lake Biwa Bridge.

$$Ro=S^{1/3}/fb^{4/3}, \tag{5.1}$$

where S is the total buoyancy flux from the South Basin $[m^4s^{-3}]$ of about 53 $km^2$, f is the Coriolis parameter, b is the width of the South Basin and the velocity scale, $(S/b)^{1/3}$ was used. A current with a small Ro is strongly deflected by the Coriolis force,

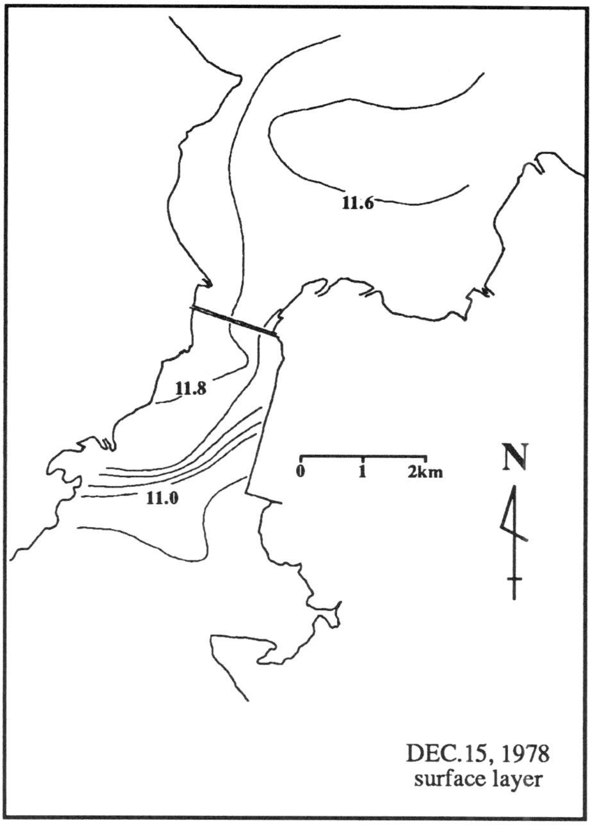


Figure 5.2. Temperature pattern in the surface layer.

so the net exchange across the LBB section is small, while currents due to strong cooling (with a large Ro) become longitudinal, resulting in a water exchange comparable to the river outflow from the South Basin. To verify such deflection of the current with the secondary component parallel to the LBB section, direct measurements of the current field were carried out. Temperature surveys were carried out several times per winter during the years 1980-1985. Each survey in the grid area in Figure 5.1 took approximately 3-4 hours. Due to the intermittent nature of the current there were times when no systematic current pattern occurred during the survey period. The temperature surveys were augmented after 1981 with continuous measurements performed with a thermistor chain and a current meter (Aanderaa TR-1 and RCM-4) at the anchor station. Later, the continuous measurements were extended to include four mooring stations (locations A to D in Figure 5.1). Surface and bottom temperatures at the stations in 1983 and 1984 are shown in Figures 5.4 and 5.5.

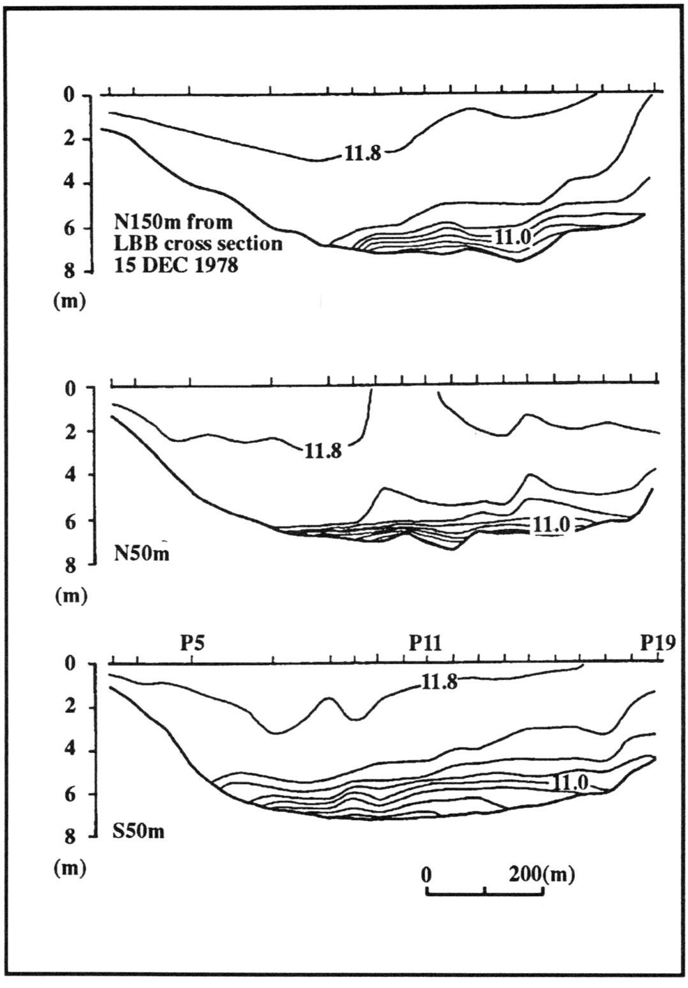


Figure 5.3. Transverse temperature distributions in the cross-sections close to Lake Biwa Bridge.

The intermittent nature of the dense bottom currents (hatched) as well as the presence of internal seiches during a nearly steady cooling of the water temperature are evident in Figures 5.4 and 5.5. Internal seiches in winter have large amplitudes because of small temperature differences, and carry cold bottom water of about 8 °C from a depth of 40 m, or greater, toward the South Basin. This sloshing rarely occurs in shallow regions, but occurs quite often in areas deeper than 15 m (stations B to D). The measuring systems did not distinguish between the return of cold water due to sloshing from the North Basin and water from the dense bottom current due to cooling of the South Basin. However, we can distinguish the dense bottom current (cooling-generated) from the periodic return current (sloshing-generated) by tracking the cooling trend from station to station.

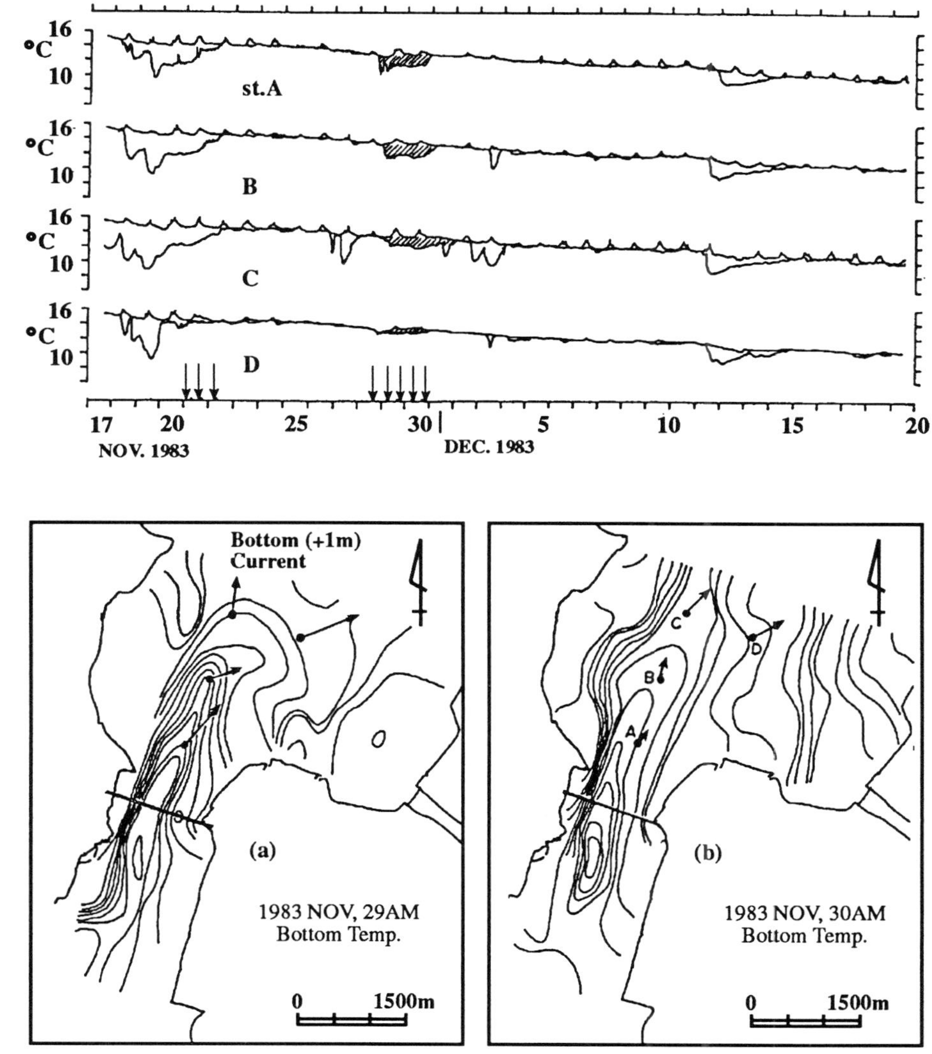


Figure 5.4. Variations in the surface and bottom temperature at the anchor stations and bottom temperature patterns when dense bottom currents are present in 1983.

The dense water underflowing into the North Basin bears similarities to currents observed in the ocean. Smith (1975) noted that currents with high friction and no entrainment flow across the contours, while currents with high mixing rates flow along the contours. Two bottom-temperature distributions from the survey are shown in the lower part of Figures 5.4 and 5.5. The current with sharply edged temperature structure flowed along the thalweg, while the current flowing along the depth contours was observed under moderate northerly wind, although it is difficult to specify the exact magnitude. The streamtube model of such flow indicates that rotation should affect the currents in Lake Biwa. As a retarding term for friction, inversely proportional to the Reynolds number, was included in the calculations, the results suggested that the path of the current was sensitive to eddy viscosities around 0.4 $cm^2\ s^{-1}$, while a value of 1 $cm^2\ s^{-1}$ had little effect. Thus, it is likely that the current was in the rotating regime, however, the mixing in the North Basin in mid-winter was so strong that the water from the South Basin would be quickly dispersed or mixed on entering the North Basin in spite of the path of the plume.

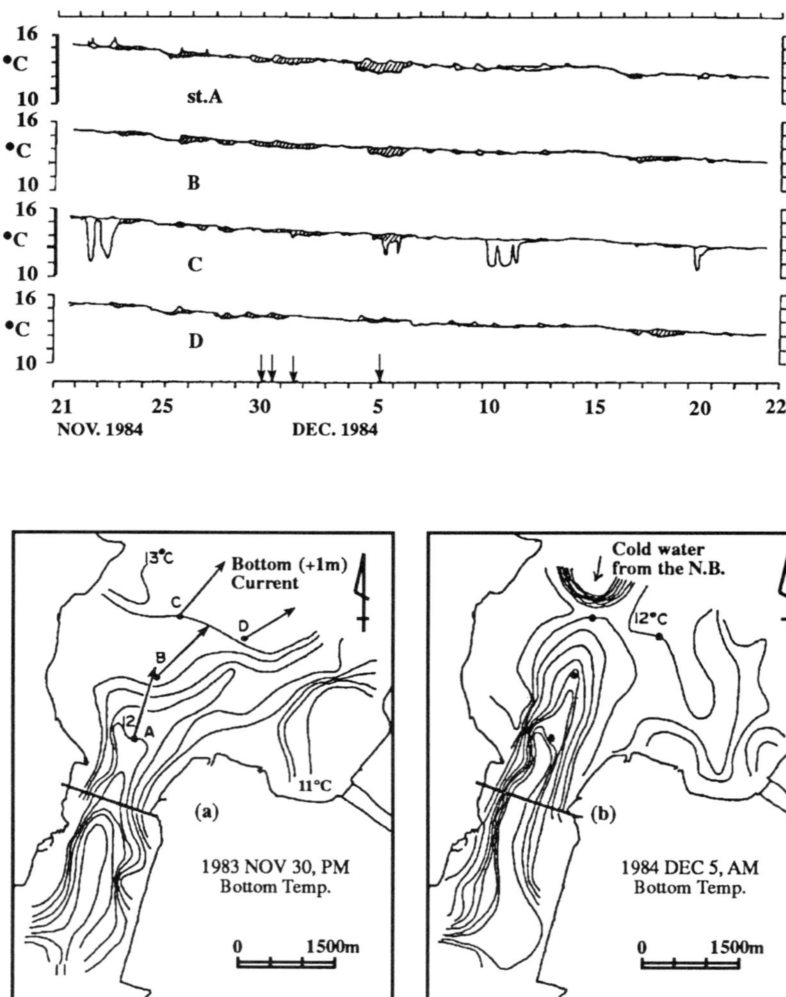


Figure 5.5. Variations in the surface and bottom temperature at the anchor stations and bottom temperature patterns when dense bottom currents are present in 1984.

In addition to determining the nature of the water exchange between the two basins, we also examined the generation of dense water in the South Basin. Studies have been made of similar processes in long bays, fjords, estuaries (Mollor, 1984; Nunes and Lennon, 1985) and sidearms of cooling ponds (Sturm and Kennedy, 1980; Monismith *et al*., 1990), however, in these rotation was not considered. For the current considered here, the wind has two opposing effects on the stratification: 1) cooling to form the current and, 2) mixing to stir up the established current. In winter, the net radiation gain was only 20-30% of the incident solar radiation, which did not exceed 200 $Wm^{-2}$ and was typically around 100 $Wm^{-2}$. As radiation was nearly in equilibrium, heat loss from the lake surface is given by conduction and evaporation $H_L = k_e(T_s - T_e)$, where $T_s$ and $T_e$ are the surface and equlibrium water temperatures, respectively. A relationship, $T_e = T_a - 0.9$ was found to hold where $T_a$ is the daily main temperature obtained by direct measurement of the water temperature of a shallow bath exposed to the atmospheric conditions at the site. The heat loss coefficient, $k_e$ [$Wm^{-2} K^{-1}$] has been found empirically to be approximately seven times the wind speed, [$ms^{-1}$], ignoring variation of $T_s$.

The observed water temperature of the dense bottom current was only little different from that calculated for the South Basin using the above formula. Given the South Basin dimensions of 15 km length and 4 km width, with a possible maximum heat loss of 700 $W/m^2$, and taking the coefficient of thermal expansion as $1.1 \times 10^{-4}\ K^{-1}$, then the Rossby number (equation 5.1) is about 0.41, which seems somewhat low. Using the minimum width at the LBB section, 1 km, Ro is estimated as 2.5.

The typical longitudinal surface temperature profile along a ship course is shown in Figure 5.6. Again, the steeper temperature front exists 2-3 km south of the LBB. Although the South Basin is not long enough to observe the equilibrium temperature, temperature differences are within about 1 °C of $T_e$ and a major temperature jump of 2-3 °C occurred a short distance from the LBB. The inset diagram shows a simplified longitudinal schematic of the current, where the water temperature in the South Basin is almost uniform and the water plunges to the bottom layer at the surface front, forming a density interface.

In the winter of 1986, two field measurement programmes were carried out in the South Basin. At Karasaki Point in the southwest the wind-driven current profiles were measured. Estimates were also made of vertical eddy diffusivity and the spiral structure of velocity profiles in the process of destratification were discussed (Okubo and Muramoto, 1987a, 1988). At the same time, profiles of the dense bottom currents were measured at the LBB section (Okubo and Muramoto, 1987b). Extracting profiles of a pure density current was difficult due to the presence of wind-driven currents.

Moreover, the LBB section is close to the node of the uninodal surface seiche and the associated oscillating current (with a period of 4 hours) often becomes predominant. Figure 5.7 shows hourly averaged current profiles at the three piers, with the vertically averaged velocities removed. The secondary current, u (positive westward), and the primary current, v (positive southward), are shown by lines and circles, respectively. The profiles taken at 1000 and 1500 on December 12 are shown; the former is a typical

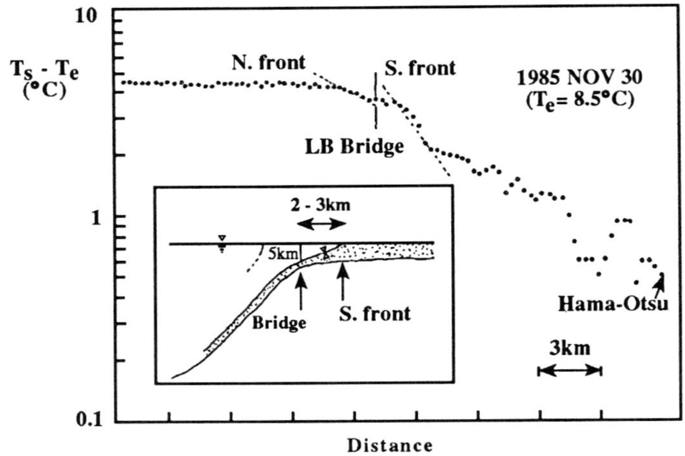


Figure 5.6. Longitudinal surface temperature profile relative to the equilibrium temperature.

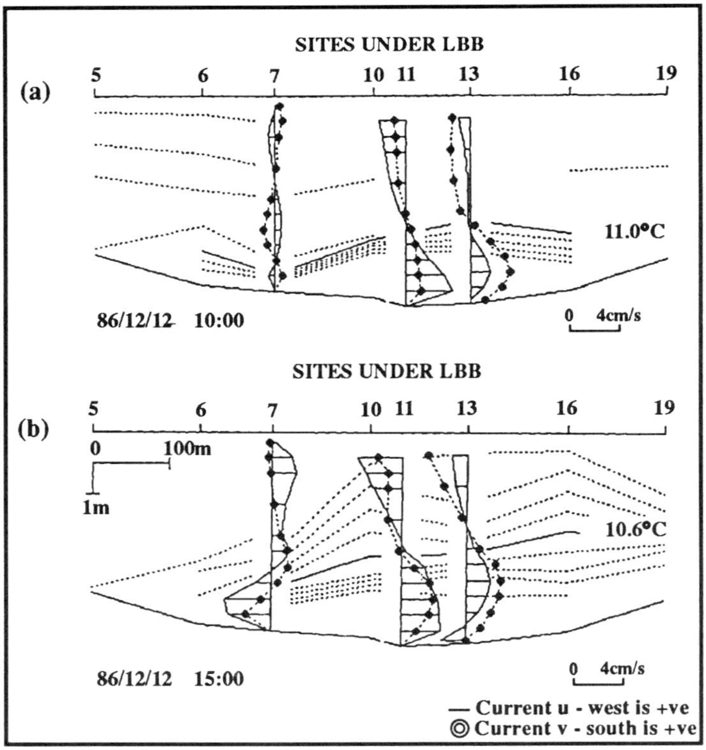


Figure 5.7. Normal and cross velocity profiles of the dense bottom current, changing to those of the currents due to northerly wind at three sites under Lake Biwa Bridge. Measurements at two times on 12 Dec 1986; (a) 10 am and (b) 3 pm when the northerly wind had developed.

density current with little effect from seiching or wind, while the latter is a stratified wind-driven current resulting from the northerly afternoon wind.

## Summary

The dense bottom current forms a temperature front in the surface layer 2-3 km south of the Lake Biwa Bridge. Water from the South Basin flows into the North Basin, extending several kilometres north of the bridge. In winter, strong winds and a spatially variable, weak thermocline cause mixing of the South Basin inflow with water in the upstream North Basin.

The Rossby number is within the range of 0.1 to 1 from both the buoyancy flux estimation and the deflection relationship obtained from current measurements. Verification was made of the Coriolis effect on the density current. The discharge is strongly affected by the rotation and the intermittent nature of current was demonstrated from the results of the field observation.

## References

Larson, R., (1986) Coriolis generated secondary currents in channels. *J. Hydr. Engrg. ASCE*, 112: 750-767.

Moller, J. S., (1984) Hydrodynamics of an arctic fjords. Series paper 34. Tech. *Univ of Denmark*.

Monismith, S., J. Imberger, J. and M. L. Morison, (1990) Convective motions in the sidearm of a small reservoir. *J. Limnol. Oceanogr.*, 35(8): 1676-1702.

Nunes, R. A. and G. W. Lennon, (1985) Episodic density current and stratification in Spencer Gulf, South Australia. *IUTAM Symposium on Mixing in Stratified Fluids*. Margaret River, Western Australia.

Okubo, K. and Y. Muramoto, (1987a) Currents in the southern basin of Lake Biwa. *Annuals Disast. Prev. Res. Inst. Kyoto Univ*. 30B-2: 569-582.(J)

Okubo, K. and Y. Muramoto, (1987b) Density currents induced by the surface cooling in Lake Biwa. Proc. 22nd Cong. IAHR, Laussanne, C1: 126-131.

Okubo, K. and Y. Muramoto, (1988) Vertical water diffusivity of wind-driven currents in a shallow lake. *Proc. 6th Cong. APD, IAHR, Kyoto*, Vol.III: 193-200.

Oonishi, Y. K. Ookubo, Y. Muramoto and A. Harashima, (1980) Coriolis effect on a buoyancy driven current. *Annuals Disat. Prev. Res. Inst. Kyoto Univ*. 23B-2: 351-371.(J)

Smith, P. C., (1975) A streamtube model for bottom boundary currents in the ocean. *Deep Sea Res.* 22: 853-873.

Sturm, T. W. and J. F. Kennedy, (1980) Heat loss from sidearms of cooling lakes. *J. Hydr Div. ASCE* 106 HY5: 783-804.

# 6

# Numerical Simulation of Density-Induced Currents Between the North and South Basins of Lake Biwa

Y. Oonishi

## Density driven currents in Lake Biwa

Lake Biwa consists of a large, deep North Basin and a smaller, shallow South Basin (Figure 6.1). The water in the South Basin has considerably different properties to that in the North Basin. The outflowing rivers from Lake Biwa are all located in the South Basin, which causes a southward current in the South Basin, so that the water from the North Basin flows into the South Basin by passing under the Lake Biwa Bridge (LBB). The total water volume flowing across this section was about $4 \times 10^9$ $m^3$ in 1977. The cross-sectional area of the lake section directly under the bridge measures 6290 $m^2$, therefore, the river flow produces a mean discharge velocity of 0.02 $ms^{-1}$ at the LBB section. Counter currents flowing from the South to the North Basin, however, driven by pressure gradients associated with surface seiches, internal seiches and lastly by buoyancy driven currents in winter, often overwhelming this river induced southward flow.

Scientific studies have been performed in Lake Biwa since the turn of this century and the surface seiche and internal seiche phenomena have been well defined. There are three major natural frequencies of the surface seiche: 240 mins, 70 mins and 30 mins (Imasato, 1971). Of these major frequencies, the first mode oscillation of the seiche has its antinode near the LBB cross-section and is important in the exchange process. The internal mode seiche can be observed in the deep North Basin from spring to autumn

Physical Processes in a Large Lake: Lake Biwa, Japan
Coastal and Estuarine Studies, Volume 48, Pages 53–64

when the water is stratified, but usually cannot be observed at the LBB cross-section. Okamoto and Yagi (1969) showed that the large amplitude oscillation associated with the internal mode seiche can be seen in the South Basin. The natural frequency of the internal seiche varies according to the magnitude of the temperature stratification. Temperature observations and numerical simulations carried out by Kanari (1974) showed that the period of the internal seiche is about two days in the summer season.

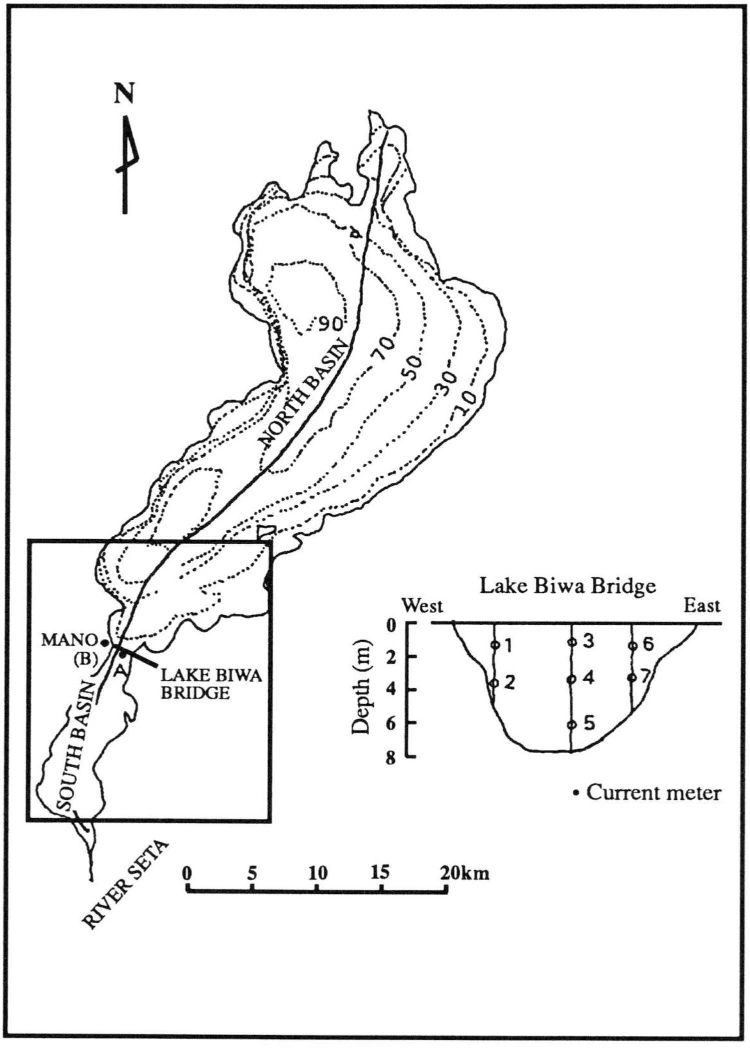


Figure 6.1. Map of Lake Biwa. The thick line drawn from north to south is the longitudinal main axis applied in the first numerical model. The rectangle is the calculation area in the second numerical model. Lake Biwa Bridge cross section is drawn with the location of seven current meters.

Muramoto *et al*. (1979) defined the structure of the water and heat budgets of the South Basin using temperature and velocity data obtained at the LBB. The Ministry of Construction maintains seven current meters with temperature sensors on three major piers of the bridge. The current and temperature data obtained in 1977 were read at 15-minute intervals. Analyzing these data, Muramoto *et al.* (1979) concluded that the mean inflow into the South Basin through the LBB cross-section was equal to the outflow from the southern end of the South Basin. That is, the effects of the local river inflow, rainfall on the basin, and evaporation, were all relatively small in the South Basin. Their analysis also showed that the heat transfer across the surface of the South Basin had the same order of magnitude as the advective heat flux through the LBB cross-section. These two components had opposite signs, and their difference was in agreement with the monthly heat gain and loss from the South Basin. The inflow through the LBB cross-section tended to reduce the amplitude of temperature variations in the South Basin.

Muramoto *et al.* (1979) also performed a spectral analysis on the heat transfer through the LBB cross-section. They separated the time series of current velocities and temperatures into seven frequency bands: (1) over 6 hours, (2) 3 to 6 hours, (3) 2 to 3 hours, (4) 1.5 to 2 hours, (5) 1.2 to 1.5 hours, (6) 1 to 1.2 hours and (7) under 1 hour. For each frequency band, heat transfer was evaluated by averaging the product of temperature and velocity. The quasi-steady current (frequency band (1), with a period over 6 hours) was divided into a cross-sectional mean value and a deviation. Finally, they isolated the shear effect of the quasi-steady current (the heat transfer due to the deviational component). The shear effect was defined as $(u^* - u)(\theta^* - \theta)S$, where u is the mean current velocity, $\theta$ is the mean water temperature, $u^*$ and $\theta^*$ are actual values over a 6 hour period of outflow and S is the cross sectional area at LBB. The analysis showed that the dispersion due to the shear effect of the quasi-steady flow was dominant in the heat transfer phenomena through the LBB cross-section, with an effect about 10 times greater than that of the total transfer by the residual components with periods less than 6 hours.

Figure 6.2 shows the separated heat transfer due to the shear effect of the quasi-steady flow. This figure demonstrates that (1) from late November to December there were intermittent heat transfers from the North to the South Basin, (2) from June to August there were also intermittent heat transfers from the South to the North Basin. In winter the water temperature is generally lower in the shallow water of the South Basin, and the warmer North Basin water flows into the south near the surface, while the surface colder and denser South Basin water counterflows into the North Basin near the bottom. The intermittent North to South Basin transfers observed in the winter may be related to phenomena associated with density current. In summer, the South Basin water is a little warmer than the North Basin water, but the difference is not large enough to explain the negative heat transfer shown in Figure 6.2. The summer heat transfer is probably caused by the mixing supported by the internal mode seiche forced by intermittent strong winds.

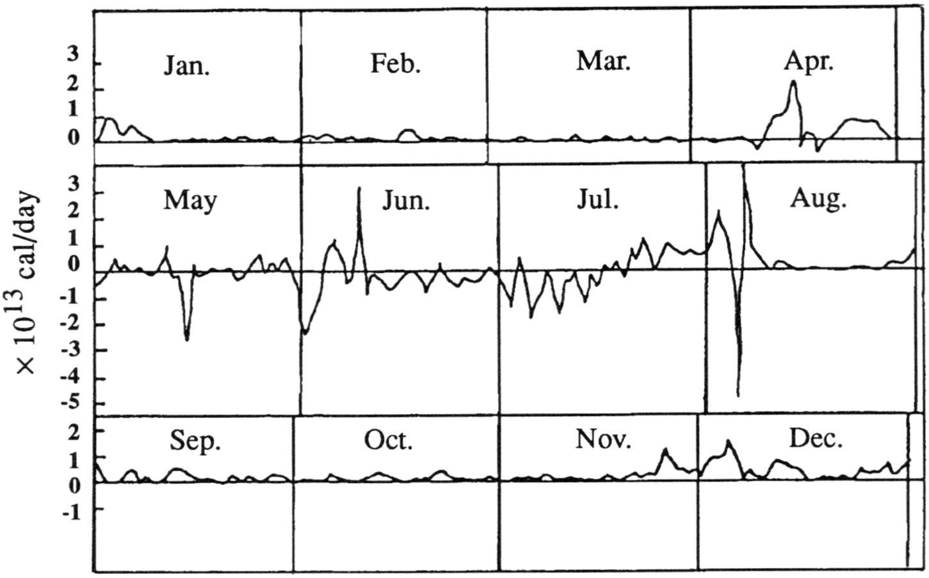


Figure 6.2. Annual variation of daily heat transfer due to shear effect of quasi-steady flow. The positive sign means heat flux toward the south.

# A two-dimensional numerical model

In order to simplify calculations the lake width was assumed to be constant and only the water motion in a two-dimensional section along the longitudinal main axis of Lake Biwa (Figure 6.1) was considered. The lake bottom profile along the section is shown in Figure 6.3 by a broken line; the computational domain and the arrangement of the non-uniform grids are shown by solid lines. It was assumed that there was a hydrostatic balance in the vertical equation of motion and the effect of the earth's rotation was neglected. In the horizontal, turbulent viscosity $\nu_h$ and diffusion coefficients for heat $K_b$ were both taken as $10^4$ cm$^2$ s$^{-1}$, and the vertical thermal diffusivity $\kappa_v$, as 0.1 cm$^2$ s$^{-1}$. The vertical eddy viscosity, $\nu_v$ , was taken as 0.3 cm$^2$ s$^{-1}$ at depths of more than 10 m, and vertical eddy viscosity changed linearly in the surface mixed layer (0 to 10 m) with a value at the lake surface of 90 cm$^2$ s$^{-1}$. The vertical profiles of the initial water temperature (a typical temperature distribution seen in August in the North Basin) and the assumed vertical eddy viscosities are shown in Figure 6.4.

Numerical calculations were performed on the premise that the water was initially at rest, the water temperature was horizontally uniform (as shown in Figure 6.4), and a south wind of 6 ms$^{-1}$ began to blow abruptly at time t = 0 and stopped at time t = 10 hours. The time integration was continued until t = 270 hours.

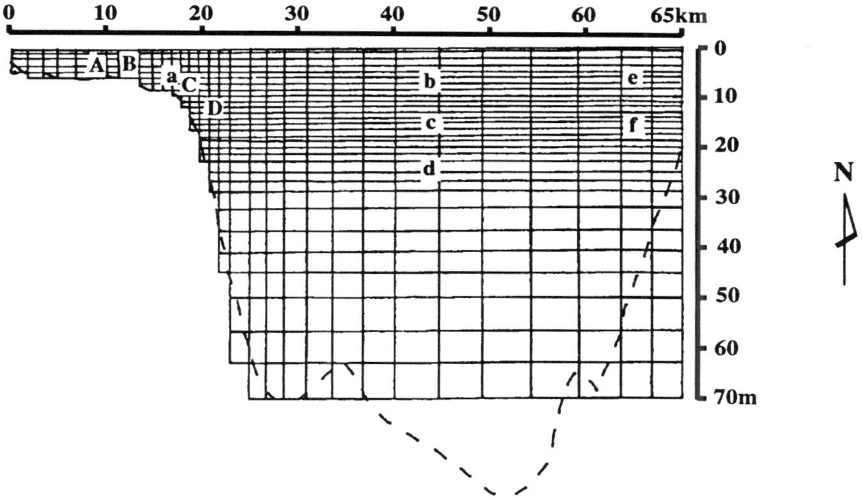


Figure 6.3. The two-dimensional section along the main axis in Figure 6.1. The broken line shows the actual bottom profile. The solid line shows the area of calculation and the arrangement of the non-uniform grids.

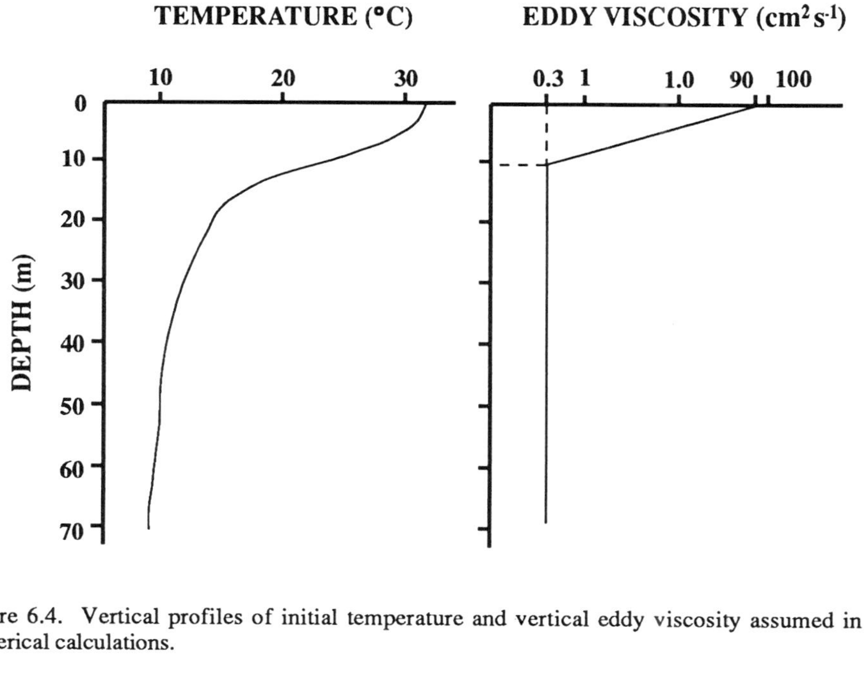


Figure 6.4. Vertical profiles of initial temperature and vertical eddy viscosity assumed in the numerical calculations.

The calculated temperature changes for several locations along the basin are given in Figure 6.5. On initiation of a south wind the warm surface water of the South Basin was driven to the north and internal seiches were induced in the North Basin. Comparing the results with field observations obtained under similar wind conditions (Figure 6.6) confirmed the findings of the numerical model; the abrupt temperature drop shown in the observation was probably induced by the internal mode seiche. Calculated changes in the horizontal current velocities over time at several points in the South Basin are shown in Figure 6.7. This figure indicates that the water motion near the LBB cross-section is dominated by the surface seiche in the South Basin (with a natural frequency of about 4 hours), while the motion in the North Basin is dominated by internal seiches. The calculated rate of the temperature decrease was smaller than that observed. This difference may have been the result of assuming a uniform lake width for our calculations.

## A three-dimensional numerical model

In winter, the dense water in the South Basin flows along the lake bottom into the North Basin (Chapter 5). A numerical simulation of the origins of this flow is now discussed.

To simulate this process, a three-dimensional dynamic model which included the Coriolis effect was used, but again it assumed a hydrostatic balance in the vertical equation of motion. The computational assumptions are shown in Table 6.1. At the northern and eastern open boundaries, the temperature and velocity conditions were set by values obtained from Kataoka's (1980) numerical model of the entire lake using a horizontal grid size of 2 km. In our model, a 500 m grid size was used to yield more precise information about the water exchange at the cross-section between the two basins. The vertical grid partition was constructed as follows: when the water depth was less than 20 m, it was divided in half; when it was between 20 m and 25 m, the top layer was 10 m deep and the lower layer was the residual; when the water depth was over 25 m, the column was divided into three layers, the first 10 m, the second from 10 m to 25 m, with the third being the residual. The vertical grid partitioning and other conditions of the calculations (Table 6.1) are identical to those of Kataoka's (1980) model.

TABLE 6.1 Computational constants for Case 1

| | |
|---|---|
| Buoyancy flux | $0.39 \times 10^{-3}$ cm$^{-1}$ sec$^{-3}$ |
| Vertical eddy thermal diffusivity | |
| between the 1st and 2nd layer | 2 cm$^2$ sec$^{-1}$ |
| between the 2nd and 3rd layer | 0.4 cm$^2$ sec$^{-1}$ |
| Vertical eddy viscosity | 1 cm$^2$ sec$^{-1}$ |
| Horizontal thermal diffusivity | $10^4$ cm$^2$ sec$^{-1}$ |
| Horizontal eddy viscosity | $10^5$ cm$^2$ sec$^{-1}$ |
| Coriolis parameter | $0.837 \times 10^{-4}$ sec$^{-1}$ |
| Gravity constant | 980 cm sec$^{-2}$ |

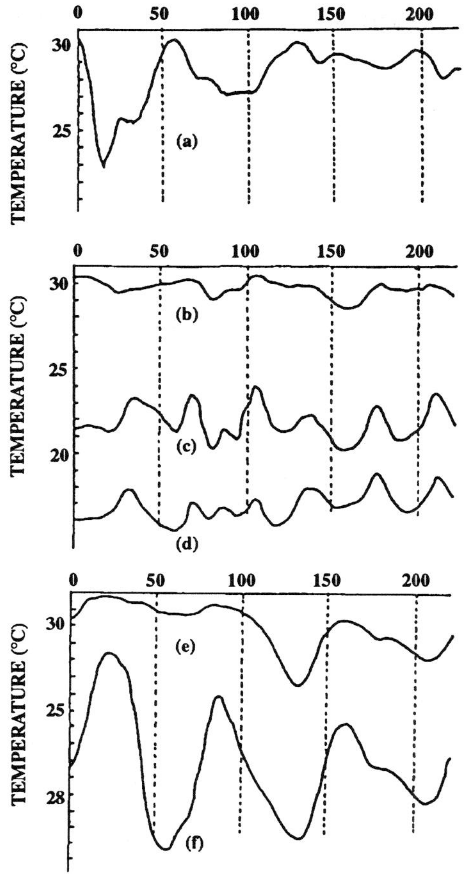


Figure 6.5. Water temperature at some locations (shown in Figure 6.3).

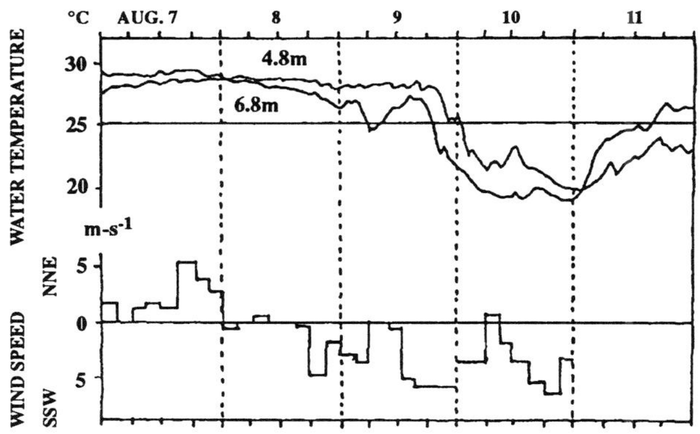


Figure 6.6. The record of water temperature at site A, and of wind speed (NNE-SSW component) at site B (Mano) in August, 1967 (after Okamoto and Yagi, 1969).

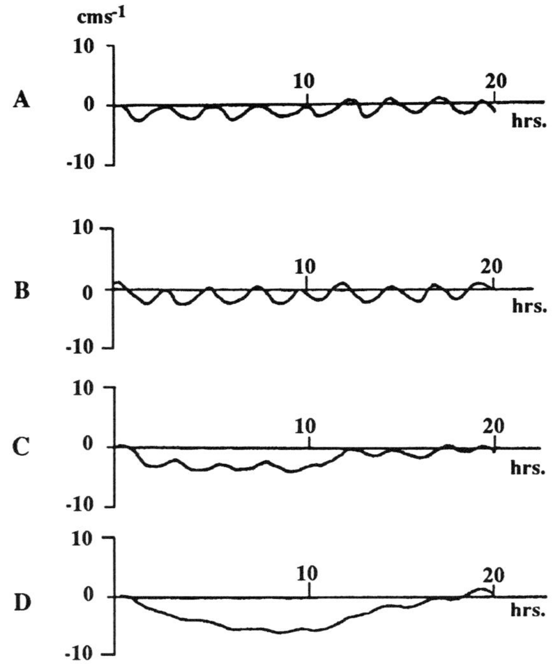


Figure 6.7. Horizontal velocity at locations (shown in Figure 6.3).

The initial conditions for the temperature distribution were taken from Kataoka's model for November 15. The initial condition for the velocity were obtained by a diagnostic method; the temperature distribution was fixed at its initial value and the corresponding steady state velocity distribution was calculated.

Calculations were performed for 5 cases. The reference Case 1 was carried out under the typical cooling conditions shown in Table 6.1. In Case 2, the buoyancy flux was doubled after t = 10 days. In Cases 3 and 4, the buoyancy flux was increased to four and eight times that of the standard case, respectively, again after t = 10 days. Case 5 was performed under the same cooling conditions as Case 1, except that the Coriolis parameter was set to zero.

A comparison between Case 1 and Case 5 at t = 30 days, illustrating the effects of Coriolis acceleration, is shown in Figure 6.8. In this figure, the current velocity in the second layer (10-25 m depth) is shown. The difference is clear. If the Coriolis effect had been omitted, a strong density current flowing from the South to the North Basin would have occurred. The Coriolis effect clearly disrupts or modifies the simple exchange flow between the basins (Condie and Ivey, 1988).

In Case 1, essentially no water exchange was evident driven by the typical cooling rate. However, in Case 3 (Figure 6.9), in which the cooling rate was increased by a factor of four, a strong density current developed at t = 15 days (5 days after the cooling rate was

**CASE 1**

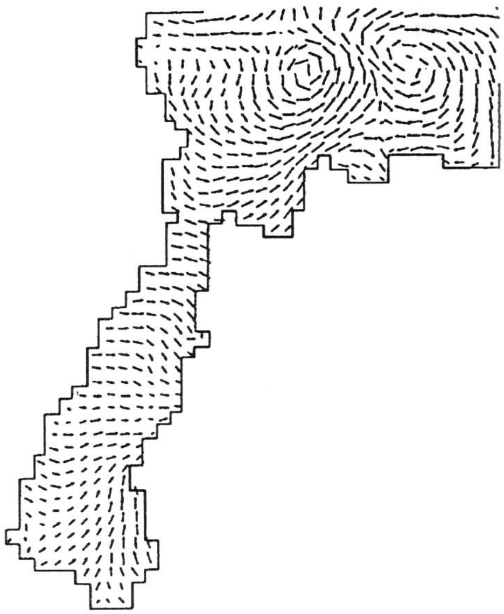

**CASE 5**

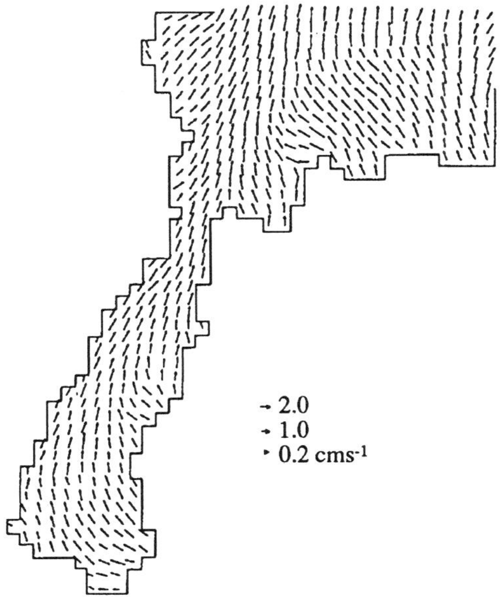


Figure 6.8. Horizontal distribution of current velocities in the second layer (10-25 m) for Case 1 and 5 (t = 30 days after start).

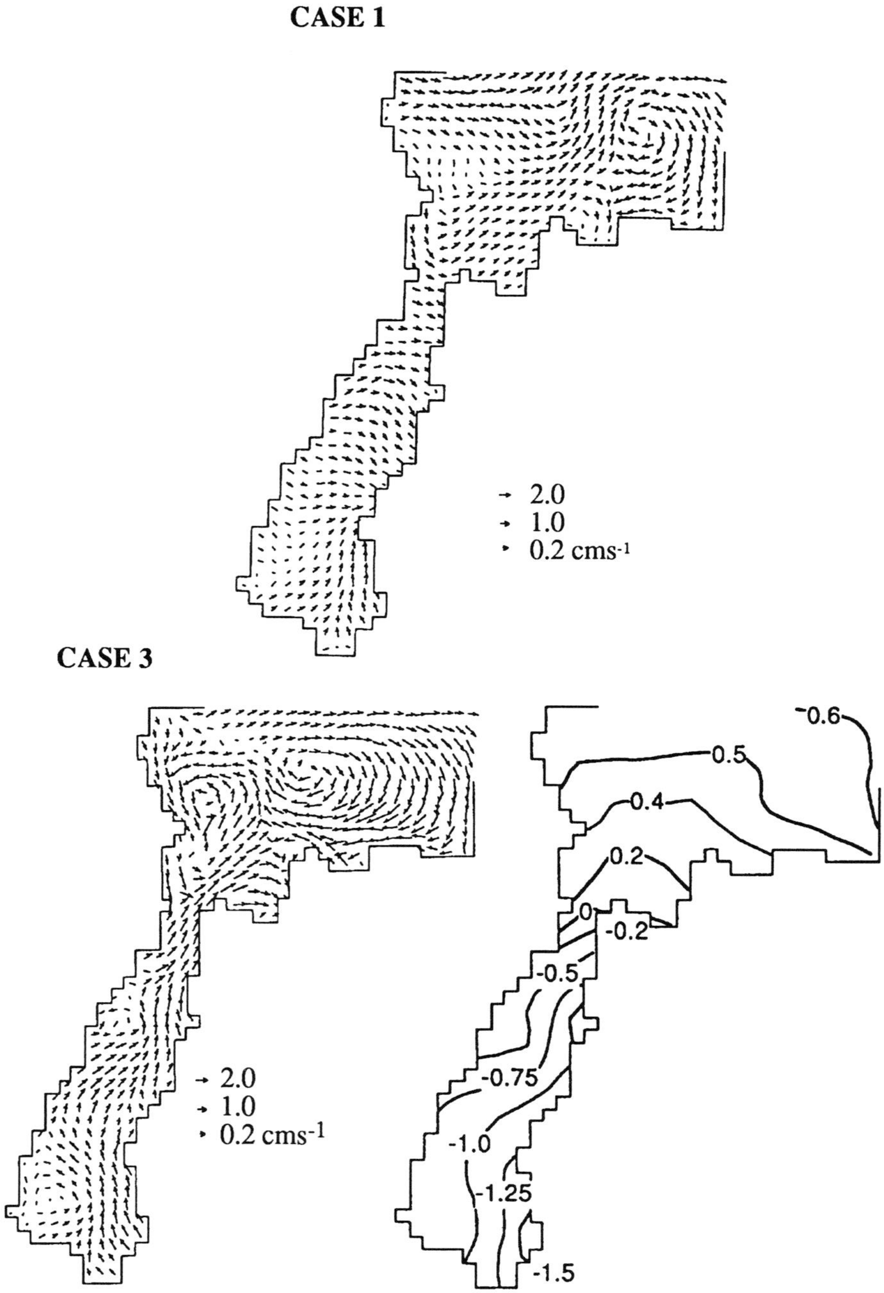


Figure 6.9. Horizontal distribution of current velocities in the second layer (10-25 m) for Case 1 and 3, and buoyancy distribution in the second layer for Case 3 (t = 15 days after start).

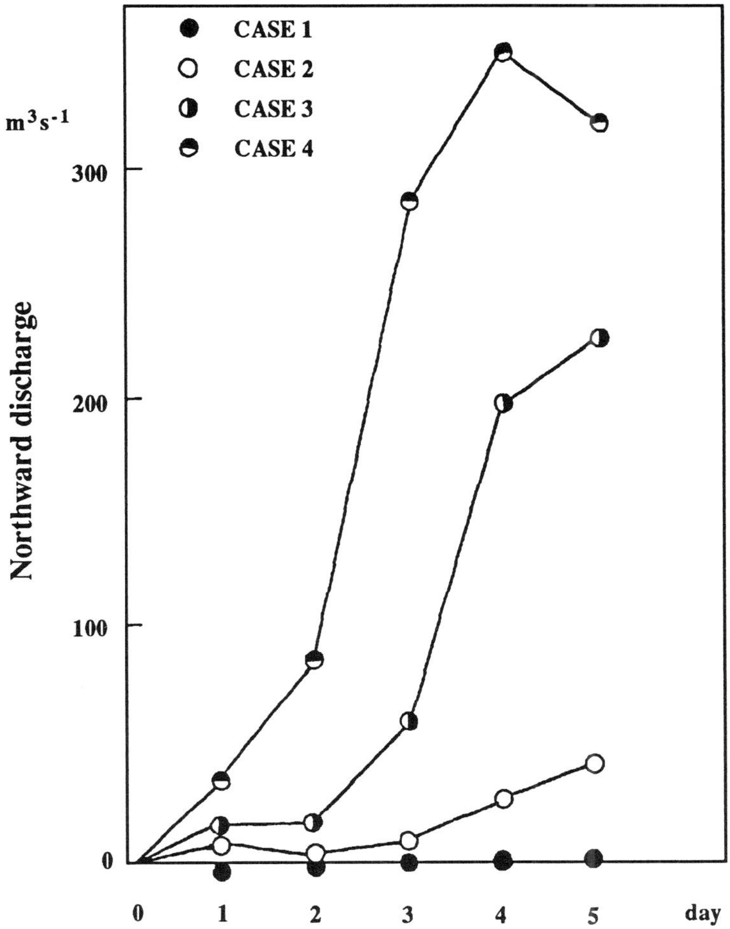


Figure 6.10. Calculated northward discharge from density currents for cases 1, 2, 3, and 4.

quadrupled). The buoyancy distribution in the second layer at t = 15 days is also shown in Figure 6.9. This figure correlates with the field observations described in Chapter 5. Due to Coriolis effects, the light North Basin water enters the South Basin along the western shore, while the dense South Basin water flows north along the eastern shore.

Finally, Figure 6.10 compares northward discharges of water due to density currents in Cases 1, 2, 3 and 4. It is shown that large exchanges of water are not induced by the typical monthly mean cooling rate, but intermittently when the cooling rate is doubled or quadrupled. This may be the reason for the large heat transfer observed intermittently in winter as shown in Figure 6.2.

# Conclusion

The heat exchange between the North and South Basin of Lake Biwa is mainly yielded by a quasi-steady component (frequency band with a period over the major frequency of the surface mode seiche) of the current. Data analysis made by Muramoto *et al.* (1979) showed the importance of a shear effect (effect by a deviation from a cross-sectional mean flow). They also found that this heat exchange due to the shear effect is an intermittent phenomenon observed in summer and winter.

Two types of numerical simulations were made. One is a vertically two-dimensional model on an internal seiche, and the other is a three-dimensional model on a density current induced by the cooling of the lake surface in winter. The former model showed that a strong wind breaks the internal seiche yielding an intermittent heat exchange as was observed in summer. The latter model showed that the typical monthly mean cooling can not entail an effective density current, but that a doubled or quadrupled cooling yields an intermittently strong density current. These two models represent two types of observed intermittence in summer and in winter.

# References

Condie, S.A. and G. N. Ivey, (1988) Convectively driven coastal currents in a rotating basin. *J. Mar Res*., 46: 473-494.

Imasato, N., (1971) Study of seiches in Lake Biwa-ko (II), *Contribution Geophys. Inst., Kyoto Univ*. 11: 77-90.

Kanari, S., (1974) On the study of numerical experiments of two layer Lake Biwa. *Jap. J. Limnol.* 35: 1-7.

Kataoka, K., (1980) On the role of the buoyancy inflow on the formation of the lake current. Graduate thesis, *Fac. Eng., Kyoto Univ*..

Muramoto, Y., Y. Oonishi, and K. Ookubo, (1979) Heat budget in the South Basin of Lake Biwa, the water exchange between the South and the North Basin of Lake Biwa at Biwako Bridge. Disaster Prev. Res. Inst. *Annuals of Kyoto Univ*. 22B2: 575-589.

Okamoto, I. and K. Yagi, (1969) Time variation of water temperature in Lake Biwa-ko (III). *Jap. J. Limnol.* 30: 108-118.

Oonishi, Y. and N. Imasato, (1974) A note on the water exchange between the South and the North Basins of Lake Biwa, *Contributions, Geophys. Inst., Kyoto Univ*. 14: 11-19.

Oonishi, Y., K. Ookubo, A. Harashima, and Y. Muramoto, (1980) Coriolis effect on a buoyancy driven current. Disaster Prev. Res. Inst. *Annuals of Kyoto Univ*. 23B2 (35):1-371.

# 7

# Spectral Irradiance and Optical Properties

R. Tsuda and M. Nakanishi

Measurements of underwater spectral irradiance are of importance when using optical methods to measure tubidity and biological primary production. In this section, we present basic spectral irradiance data from measurements made with a multichannel instrument (Biospherical Instr., MER-1000; Kishino *et al.*, 1984) at various depths in both the North and South Basin of Lake Biwa.

## Light Attenuation and Spectral Distribution of Downward Irradiance in the North and South basins

Light attenuation with increasing depth in natural waters is caused by a combination of absorption and scattering, mainly due to the presence of suspended particles, dissolved substances, and the clarity of the water itself. In typical inland and coastal waters where excessive suspended particles and dissolved substances exist, underwater irradiance is rapidly attenuated with depth below the free surface.

Figure 7.1 demonstrates the relative decrease in total irradiance with increased depth in the South and North Basin at various stations shown in Figure 7.2. Irradiance in the South Basin is attenuated rapidly due to high concentrations of suspended particles and or resuspended sediments and dissolved substances. At Station 1 in the South Basin, the euphotic depth $Z_{1\%}$ (defined as the depth to which 1% of incident light penetrates) is about 3.0 m. In June 1981 at another South Basin location (Station 3), very turbid water was observed in layers from 2 m downward near the lake bottom. The Chl-*a* concentration in this layer was 1.7 mg/m$^3$ which was lower than the value of 4.1 - 4.6 mg/m$^3$ obtained in March 1981. However, the concentration of suspended particles in this layer at 4.1 - 4.7 mg/m$^3$ was higher compared to the 2.2 - 2.4 mg/m$^3$ value of March

Physical Processes in a Large Lake: Lake Biwa, Japan
Coastal and Estuarine Studies, Volume 48, Pages 65–76

1981. Thus, the higher attenuation in this surface layer depended mainly on the detritus concentration. The euphotic depth at Station 7 in the North Basin was about 12.0 m in March 1982 and 13.0 m in June 1981 and the results indicate that the higher concentrations of suspended particles and dissolved substances found in the waters of the South Basin are responsible for these differences in the euphotic depth.

Seasonal changes in the relative decrease of total irradiance in the North Basin are shown in Figure 7.3. The euphotic depth varies from 10 m to about 17 m. High attenuation of light intensity was observed in June, October, and December 1983, and at depths between 1 m and 3 m in May 1984. High light attenuation coincided with high Chl-*a* concentrations. In May 1984, the rapid decrease of light between 1 and 3 m reflected the presence of a dense bloom of fresh water red tide called *Uroglena americana* which frequently forms in May and or the beginning of June in Lake Biwa. The amount of chlorophyll at the euphotic depth in December 1983 was lower (55 $mg/m^2$) than in October 1983, with light attenuation substantially diminished compared to that measured in October 1983. This seems to be attributable to the absorption by Chl-*a* in addition to the prevailing existence of accessory pigments of the blue green algae, *Aphanothece clathrata*, within the euphotic layer. The maximum transmittance of irradiance in the North Basin is found at a wavelength ranging between 540 and 570 nm as described in the section on underwater spectral irradiance. The absorption spectrum of *A. clathrata* allows efficient utilization of the radiant flux. Hence, the properties of this species would account for the high light attenuation of December.

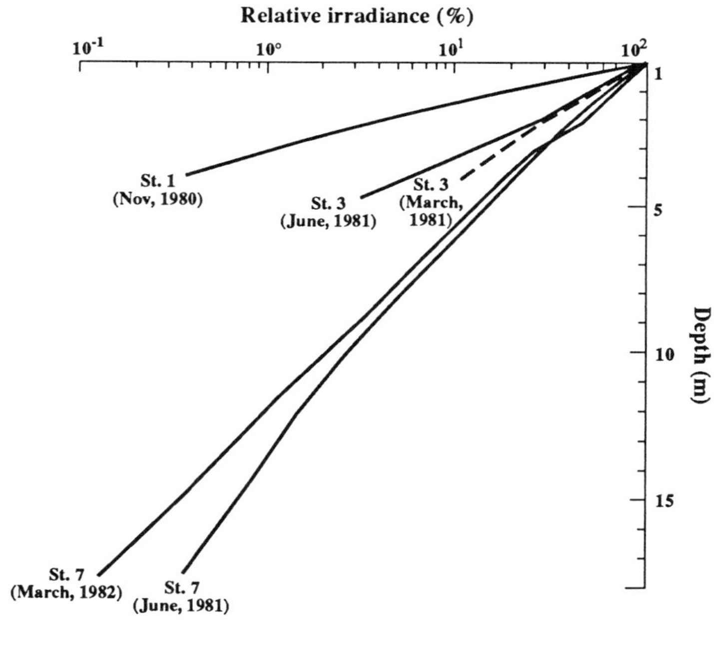


Figure 7.1. Relative irradiance at stations in the South and North Basins. Station locations are shown in Figure 7.2.

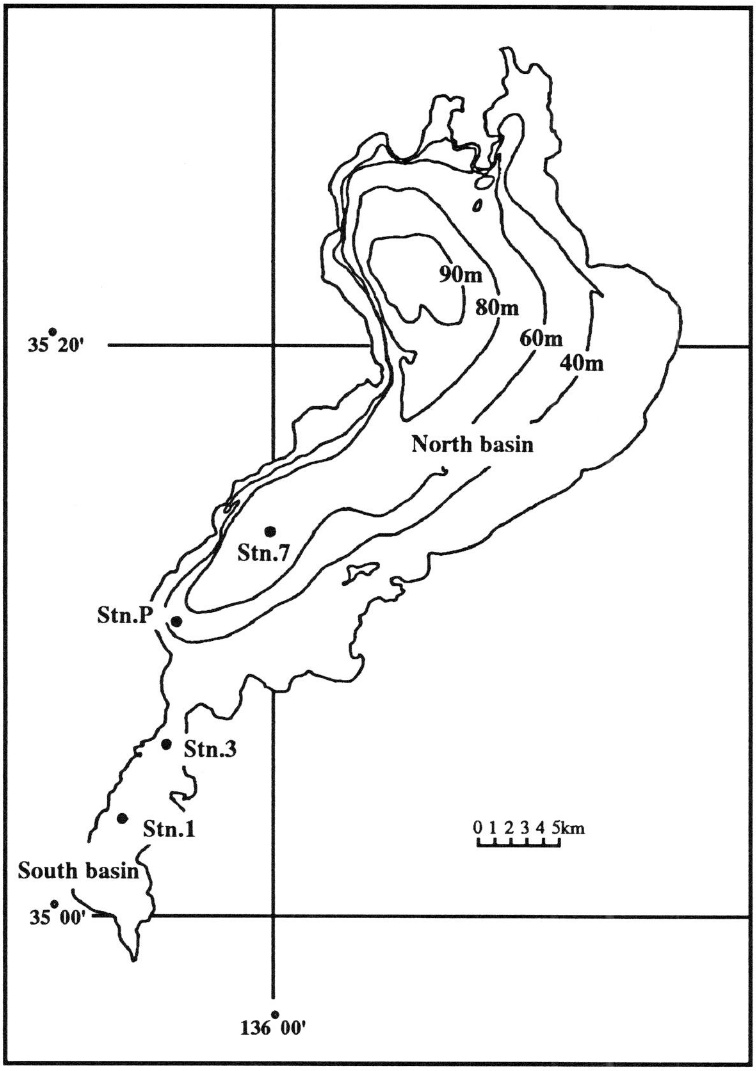


Figure 7.2. Locations of stations where optical measurements were made.

Figure 7.4 shows the spectral distribution of downward irradiance measured in the South Basin. Spectral distribution is influenced by changes in both the quantity and composition of photosynthetic pigments. A typical spectral distribution of irradiance for very turbid water observed at Station 1 shows that spectral distribution peaked at 570 nm as a result of very strong attenuation attributable to absorption by suspended particles

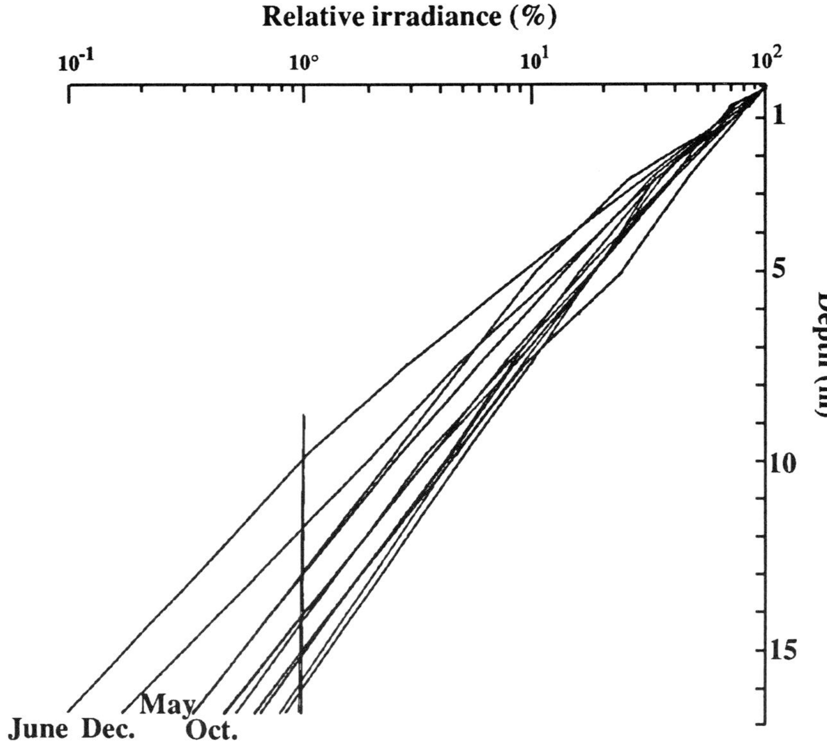


Figure 7.3. Relative irradiance on experimental days at station P in the North Basin (Tsuda and Nakanishi, 1990).

and dissolved substances in the blue region of the spectrum, while the shoulders appearing at 625 nm and 671 nm between 2 m and 3.8 m were attributable to absorption by phycocyanin and Chl-*a* (Yentsch, 1980). The maximum transmittance of irradiance at Station 3 in March 1981 was found within the 540 - 570 nm band and absorption by the blue green algal phycocyanin also occurred at a wavelength of 625 nm. With increasing depth the spectral distribution at Station 3 in June 1981 revealed a rapid attenuation in blue light. This rapid decrease in the short-wave region of the spectrum is due to absorption by both detritus and dissolved substances.

Figure 7.5 shows the spectral distribution of downward irradiance at Station 7 and other measurements taken in the North Basin. The maximum transmittances of irradiance in June 1981 and March 1982 occurred around the green part of the spectrum (540 - 570 nm), though there were some differences in the spectral distribution. In June 1981, spectral absorption occurring at 625 and 671 nm was due to accessory pigments and Chl-*a*, respectively. However, in March 1982, there was no evidence of strong absorption at 671 nm indicating a change of algal composition.

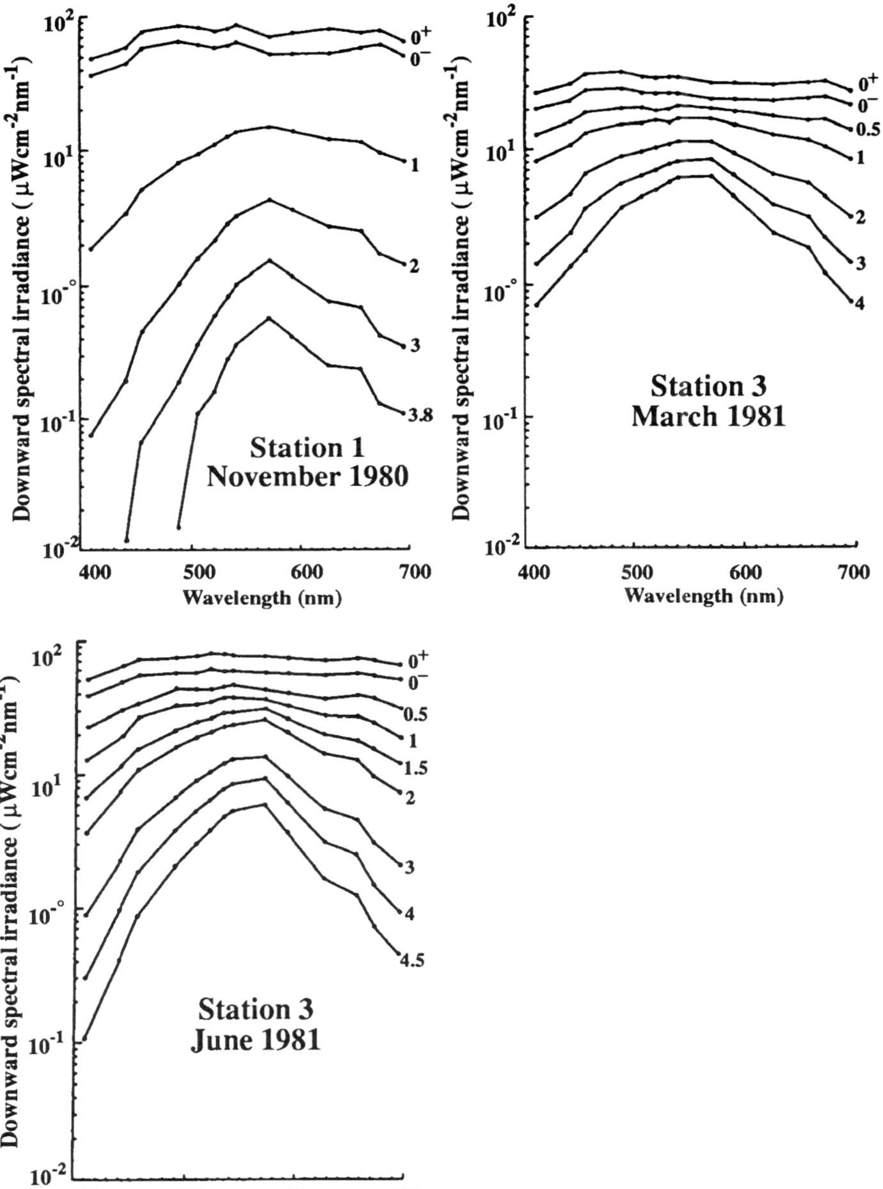


Figure 7.4. Representative downward spectral irradiance in the South Basin for a range of depths (m) at each station.

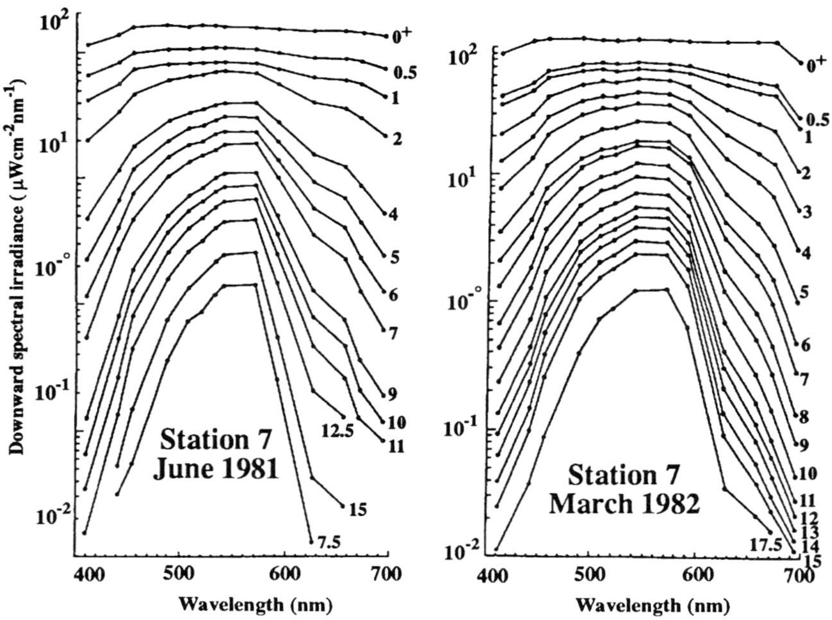


Figure 7.5. Downward spectral irradiance at Station 7 in the North Basin for a range of depths (m).

The seasonal characteristics of the spectral measurements in the North Basin of Lake Biwa can be classified into four types (Figure 7.6). In June 1983, the amount of Chl-a within the euphotic layer was about 82 mg/m$^2$ at an euphotic depth of nearly 10 m. The maximum transmittance of irradiance was found in a wavelength range of 540 to 570 nm. This spectral distribution type was accompanied by the presence of relatively high Chl-a concentrations within the euphotic layer, thus the shoulders appearing at 441 and 671 nm are attributable to Chl-a pigments. The curves for May 25, 1984 were skewed with stronger absorption of red than blue light and peaking at 570 nm. The minima and shoulders appearing at 671 nm were due to strong absorption by Chl-a. High Chl-a concentrations in the surface layer were caused by the presence of *U. americana*, concentrating around a depth of 2.5 m at a level of 17 mg/m$^3$. In March 1984, spectral irradiance distribution peaked at a wavelength of about 540 nm. This spectral distribution is frequently observed in the North Basin when the amounts of chlorophyll within the euphotic layer are relatively low (less than about 40 mg/m$^2$). The euphotic depth was between 15 m and 20 m. On the other hand, the curves for December 1983 peaked at 541 nm despite the relatively high amount of Chl-a (55 - 77 mg/m$^2$) within the euphotic layer. The plankton community within the euphotic layer was dominated by the blue green alga *A. clathrata*, which has maximum absorption spectra around 560 nm, 670 nm, and in the band between 400 nm and 500 nm (Tsuji *et al.*, 1986). A strong absorption spectrum around 560 nm resulted in a shift of the peak to the 540 nm region. These results indicate that the spectral distribution of irradiance cannot be explained by the Chl-a concentration alone.

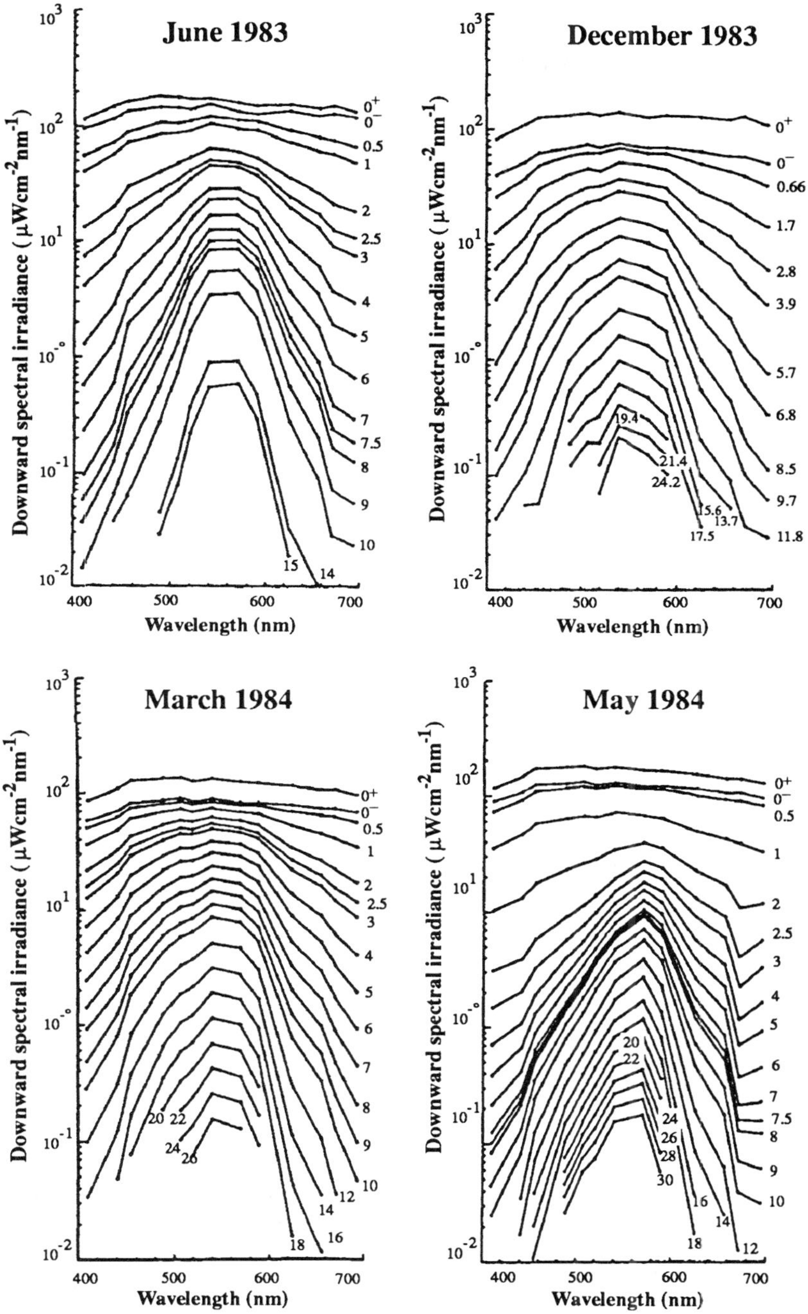


Figure 7.6. Representative downward spectral irradiance in the North Basin for a range of depths (m) (Tsuda and Nakanishi, 1990).

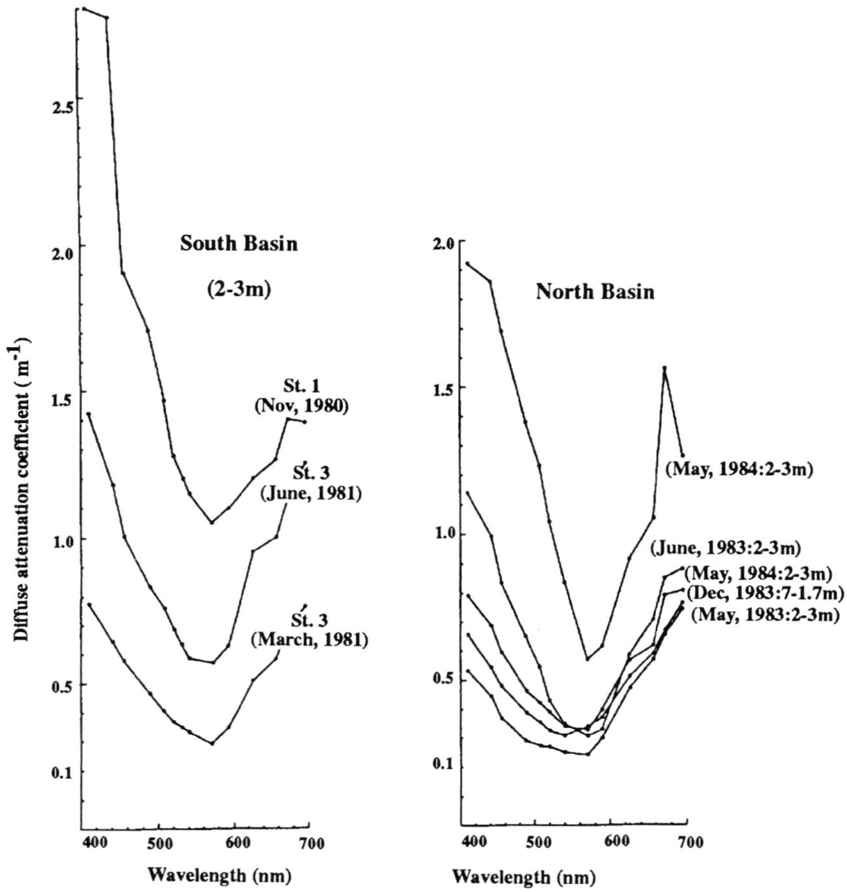


Figure 7.7. Representative curves of spectral values of diffuse attenuation coefficient K ($m^{-1}$). (Tsuda and Nakanishi, 1990).

# Optical Properties of Water in the Surface Layers of the South and North basins

The presence of suspended particles and dissolved substances in water increases absorption and scattering, resulting in a variation in the diffuse attenuation coefficient with the optical properties of the suspended particles and dissolved substances. Moreover, depending on whether the waters are clear or turbid, the shape of the diffuse attenuation coefficient K ($m^{-1}$) as a function of defined wavelength, exhibits different shapes (Morel, 1982), where K is calculated using a layer of finite thickness between two measured depths of downward spectral irradiance. Figure 7.7 shows representative K-curves in the South and North Basins.

The minimum K in the South Basin is found at a wavelength of 570 nm, but the shape of the K-curve at different times was quite different. In June 1981 (Figure 7.7), there was no sharp notch in K values mainly due to high concentrations of detritus and the curve was nearly "U"-shaped (Morel, 1982). For coastal waters, "U"-shaped curves remain flat throughout the spectral band, although the broad minimum passes from 460 nm to 560 nm with increasing detritus. However, strong attenuation is found in the blue-green light region. Light absorption in the blue region of the spectrum depends significantly on the quantity of dissolved substances which are the main cause of light attenuation in inland waters.

In March 1981 (Figure 7.7), the Chl-a concentration was relatively high (4.1 - 5.6 $mg/m^3$) and the K-curve was "V"-shaped (Morel, 1982). As Chl-a concentrations increase, the "V"-shape becomes more pronounced (November, 1980) with the effects of Chl-a absorption peaking at 671 nm. Strong absorption in the 400 - 540 nm region in November 1980 (Figure 7.7) was attributable to dissolved substances. Thus, spectral values for the diffuse attenuation coefficient in the South Basin are highly dependent on the concentration of dissolved substances.

The K-curve in the North Basin may be classified as either "U"- or "V"-shaped. When Chl-a concentrations are low, the K-curve is nearly "U"-shaped (May 1983), though absorption in the 400 - 500 nm region is not as sharp as in the South Basin due to the weak contribution of dissolved substances to total attenuation. As Chl-a concentrations increase, however, the K-curve gradually becomes "V"-shaped and the minimum of K is found at a wavelength of 570 nm. Absorption of dissolved substances also appears progressively more pronounced with increases of K resulting in a narrower K-curve in May, 1984. The K-curve in December 1983 is approximately "V" -shaped, though the minimum of K is reached at a wavelength of 540 nm, with intense light absorption by accessory pigments at around 570nm.

## Optical Properties and Biological Parameters of the Euphotic Layer

The euphotic depth varies considerably during the year in Lake Biwa due to its dependence on the concentration of suspended particles and can be related to the light attenuation coefficient. Figure 7.8 shows the relationship between the $Z_{1\%}$ depth and the averaged diffuse attenuation coefficient ($KZ_{1\%}$) or the averaged beam attenuation coefficient ($\alpha Z_{1\%}$) within the euphotic depth in the North Basin. As the K-value increases, the $Z_{1\%}$ depth uniformly decreases. The $Z_{1\%}$ depth in the North Basin can be determined as a function of the light attenuation coefficients as follows:

$$K_{Z1\%} = 8.1\, Z_{1\%}^{-1.2} \qquad (r^2 = 0.896, N=21) \qquad (7.1)$$

$$\alpha_{Z1\%}(667) = 99.7\, Z_{1\%}^{1.7} \qquad (r^2 = 0.938, N=21) \qquad (7.2)$$

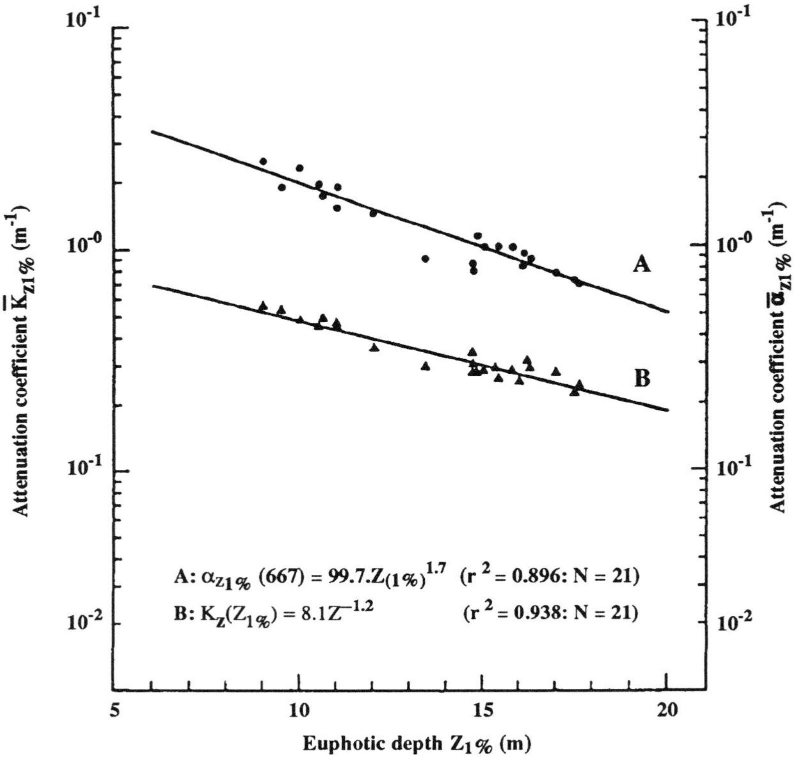


Figure 7.8. Relationship between the euphotic depth and attenuation coefficient within the euphotic layer in the North Basin. A: diffuse attenuation coefficient; B: beam attenuation coefficient. (Tsuda and Nakanishi, 1990.)

With offshore water, the thickness of the euphotic layer depends on the averaged pigment content (Lorenzen, 1972) $\alpha\ Z_{1\%}$ (667) = 99.7 $Z_{1\%}{}^{1.7}$. If thickness can be determined as a function of the amount of Chl-a within the euphotic layer, the optical properties are dependent on biogenous material. Figure 7.9 shows the relationship between the amount of Chl-a within the euphotic layer and the euphotic depth, as calculated from measurements of underwater spectral irradiance in the North Basin. When the amount of Chl-a is more than 50 mg/m$^2$, the thickness of the euphotic depth follows Lorenzen's equation, and with increasing euphotic depth, the data depart from the curve. The $Z_{1\%}$ figures of the stations in the South Basin were substantially lower from those predicted by Lorenzen's equation, indicating the presence of high concentrations of resuspended sediment.

The data of Figures 7.8 and 7.9 reveal that high amounts of Chl-a (>50 mg/m$^2$ ) within the euphotic layer are reflected in the $KZ_{1\%}$ values. When plotting the averaged Chl-a concentrations per euphotic depth against $KZ_{1\%}$ values, a strong correlation is obtained (Tsuda and Nakanishi, 1990 ). The values of this gradient can be used to derive the specific attenuation coefficient, which is important in understanding the photosynthetic efficiency of phytoplankton in utilizing light energy. This coefficient varies with algal composition (Shanz, 1985), cell size, shape (Kirk, 1982 ), physiological state, and changes in the spectral distribution of underwater irradiance (Dubinsky, 1980).

# Conclusion

The thickness of the euphotic layer in Lake Biwa does not always depend on the concentration of biogenous material. Euphotic depths in the South Basin, in particular, reflect the presence of detritus and or dissolved substances.

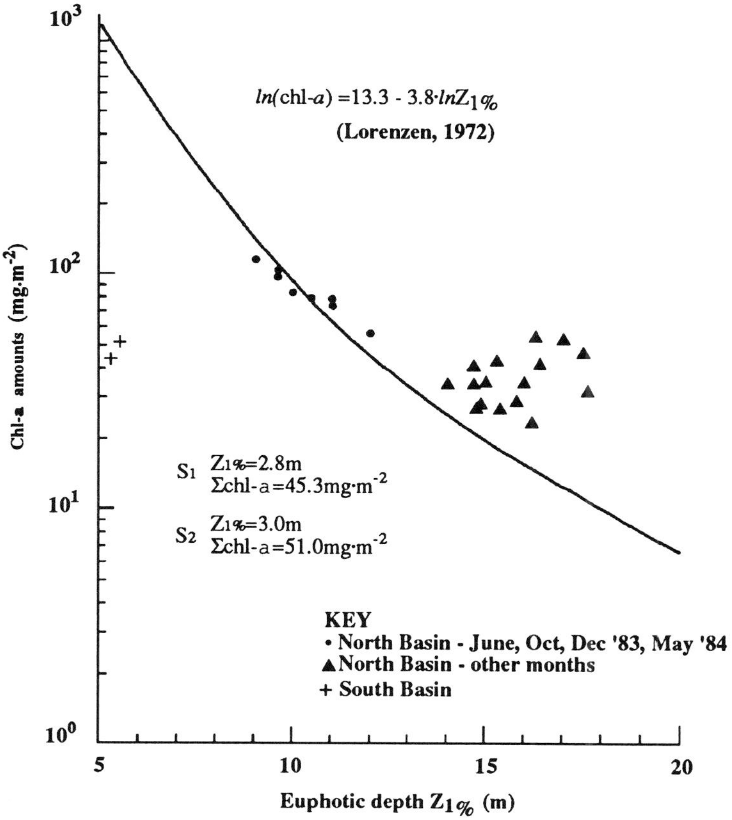


Figure 7.9. Relationship between the euphotic depth and the Chlorophyll a amounts within the euphotic depth derived from Lorenzen's statistical expression (1972), Circles: June, October, December, 1983 and May 1984; triangles: months except those aforementioned; crosses: Stations of South Basin.

# References


Dubinsky, Z., (1980) Light utilization efficiency in natural phytoplankton communities. -In *Primary productivity in the sea edited* by P. G. Falkowski,, pp.83-97. Plenum Publ. Co. New York.

Kirk, J. T. O., (1982) *Light and photosynthesis in aquatic ecosystems*, pp.1-401. Cambridge.

Kishino, M., C.R. Booth and N.Okami, (1984) Underwater radiant energy absorbed by phytoplankton, detritus, dissolved organic matter, and pure water. *Limnol. Oceanogr*. 29: 340-349.

Lorenzen, C.J., (1972) Extinction of light in the ocean by phytoplankton. *J. Cons. int Explor. Mer*, 34: 262-267.

Morel, A., (1982) Optical properties and radiant energy in the waters of the Gunea Dome and the Mauritanian upwelling area in relation to primary production. - In *The Canary Current : Studies of an Upwelling Systems*:, edited by G. Hempel, p.94-107. Rapp. p. -v. Reun. Cons. int Explor. Mer, 180. Copenhague K Denemark.

Shanz, F., (1985) Vertical light attenuation and phytoplankton development in Lake Zurich. *Limnol. Oceanogr.*, 28: 833-846.

Tsuda, R. and M. Nakanishi, (1990) Spectral distribution characteristics in downwelling irradiance and optical properties within the euphotic layer in the North Basin of Lake Biwa. *Arch. Hydrobiol*. (119: 279-292.)

Tsuji, T., K. Ohoki and Y. Fujita, (1986) Determination of photosynthetic pigment composition in an individual phytoplankton cell in seas and lakes using fluorescence emitted from picophytoplankton cells. *Mar. Biol.*, 93: 343-349.

Yentsch, C. R., (1980) Light attenuation and phytoplankton photosynthesis. In *The Physiological Ecology of Phytoplankton* ,edited by I. Morris, pp.95-127. Study in Ecology, 7. Blackwell Scientific. Publ.

# 8

# Turbidity in the North and South Basins

R. Tsuda

## Introduction

The measurement of turbidity in Lake Biwa is very important in determining transport mechanisms of contaminants and for studies of water quality. Optical monitoring of particle concentrations, including those in river plumes and bottom nepheloid layers, show that particle characteristics can vary significantly, and a detailed understanding of this variability is vital in determining light attenuation values.

Turbidity in Lake Biwa has been inferred from optical measurements (Tsuda, 1980; Tsuda *et al.*, 1990; Tsuda and Nakanishi, 1990), from dynamical considerations (Ookubo and Muramoto 1986; Okuda *et al.*, 1986), and has been further investigated by utilizing satellite images (Oonishi, 1984; Kumagai *et al.*, 1986; Kumagai *et al.*, 1990). It is well known that the water of the South Basin is more turbid than that of the North Basin and Figure 8.1 shows the surface distribution of the light attenuation coefficients commonly used to estimate the volume concentration of particles in Lake Biwa. Attenuation coefficient values in the South Basin are > $3.5\ m^{-1}$, about two times higher than values obtained from the North Basin. In this section, basic beam transmittance data measured in both the North and South Basin are presented, and the characteristics of the turbidity distribution of both basins are described.

## Analysis of turbidity

The concentration of suspended materials in Lake Biwa is dependent on both, allochthonous materials from river run-off, and autochthonous materials consisting of a

Physical Processes in a Large Lake: Lake Biwa, Japan
Coastal and Estuarine Studies, Volume 48, Pages 77–85

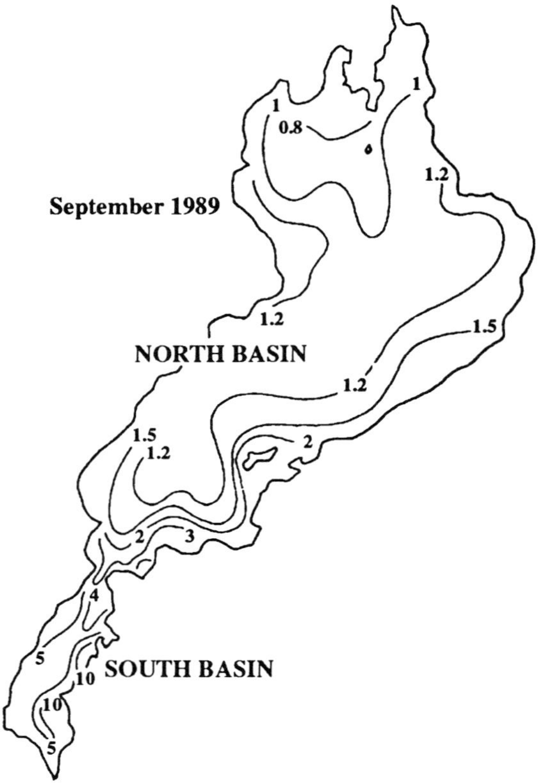


Figure 8.1. Surface distribution of beam attenuation coefficients in Lake Biwa.

quatic biological organisms and their detritus. The light attenuation coefficient, as measured from a beam transmissometer, is commonly used as an indicator of the mass or volume concentration of particles.

The beam attenuation coefficient c ($m^{-1}$) is

$$c(\lambda) = -(\ln T(\lambda))\ 1/r \tag{8.1}$$

with $T(\lambda)$ as the beam transmittance, r as the path length (0.5 m) and $\lambda$ as wavelength (667 nm). Measurements of beam transmittance made at 667 nm will yield values for the beam attenuation coefficient as follows:

$$c(667) = c_w(667) + c_p(667) + a_y(667) \tag{8.2}$$

where $c_w(667)$ is the beam attenuation coefficient for pure water, $c_p(667)$ is the attenuation coefficient for suspended particles, and $a_y(667)$ is the absorption coefficient for dissolved substances. The absolute value of $c_w(667)$ is not precisely known, but estimates range from 0.25 - 0.50 ($m^{-1}$) (Jerlov, 1976; Smith and Baker, 1981).

Dissolved substances in the water cause greatly reduced attenuation in the long wavelengths, therefore attenuation for dissolved substances is ignored,

$$c(667) = c_w(667) + c_p(667) \tag{8.3}$$

The transmissometer described in this section has been calibrated to yield a value of $c_w$ of 0.41 $m^{-1}$ (Smith and Baker, 1981). Thus from (8.3)

$$c_p(667) = c(667) - 0.41 \tag{8.4}$$

and turbidity can be defined using the values calculated from (8.4).

## Horizontal Distribution of Turbidity in the South Basin

The surface water in the South Basin is characterized by high attenuation coefficients associated with turbidity, shown in Figure 8.1. Figure 8.2 shows the horizontal distribution of beam attenuation in the near-surface water of the South Basin in May 1986. High attenuation along the east side with low values of attenuation coefficient extending from the north to the south in the shape of a tongue is the general pattern. This type of east-west gradient has also been described by Kumagai *et al*. (1986) and is also frequently observed in areas with high Chl-a concentrations in the South Basin (Nakanishi *et al*., 1986). Highly turbid water along the east side may reflect high concentrations of biogenous particles.

Values of attenuation coefficient at a depth of 2 m are relatively low compared to that at the surface, but the horizontal distribution remains similar to that found at the surface. At this depth, turbid water is also found in a zone around the south-west side where the water movement stagnates. This high attenuation coefficient is a result of accumulated suspended particles. At 3 m the horizontal distribution of turbidity has a similar pattern compared to that at the surface and at 2.0 m, resulting in beam attenuation coefficient values of 2.0 $m^{-1}$ over much of the water body in the South Basin. A turbid layer near the bottom is often observed during seasons with overall warming trends (Tsuda and Nakanishi, 1990), and in particular, the water body near Akanoi Bay shows a higher value, in excess of 4.0 $m^{-1}$.

The high turbidity water along the east side is probably due to the combined effects of the hydrological conditions, the topography of the South Basin, and the effect of human activity. In general, water on the west side of the South Basin moves towards the south and on the ease side it moves towards the north (Okamoto, 1964). The west side is directly influenced by water with low concentrations of suspended particles originating from the North Basin, while the water flowing along the east side is hardly influenced at all by the water from the North Basin. Furthermore, the topography of the lake bottom along the east side is relatively shallow and gentle and this enhances abundant upward flux of bottom sediment resuspended by wind-generated turbulence. Regular dredging along the east side further increases the concentration of suspended particles.

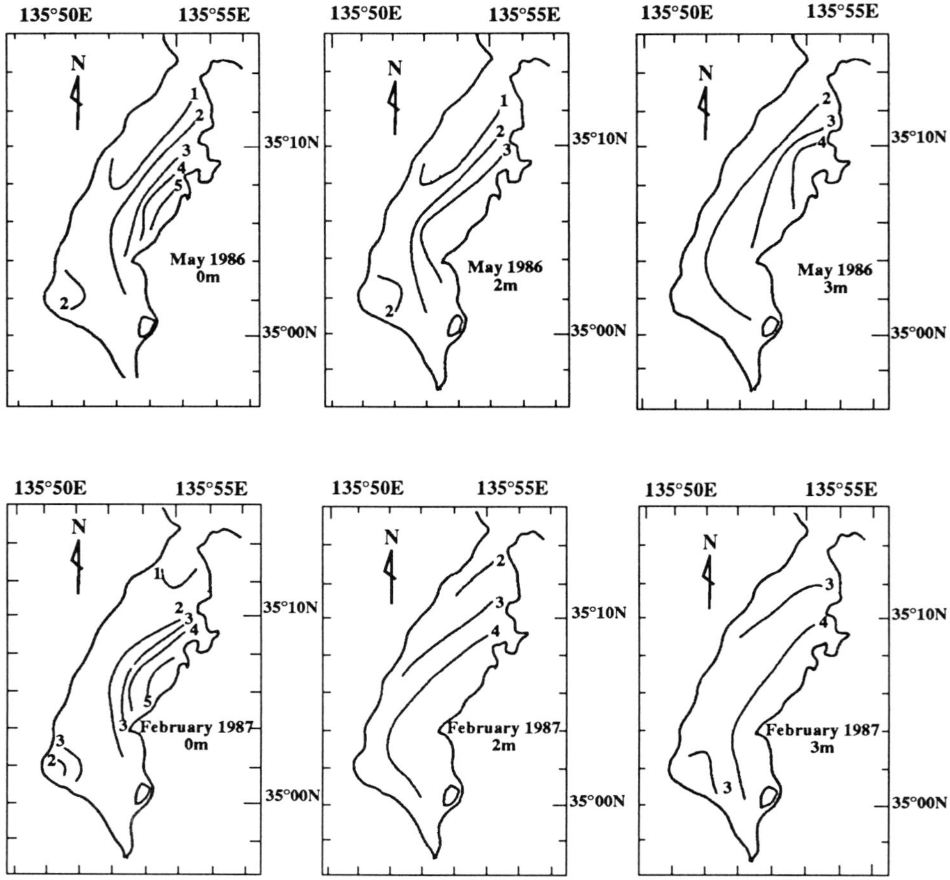


Figure 8.2. Horizontal distribution of beam attenuation coefficient ($m^{-1}$) in the South Basin in May and February 1987.

It is concluded that the combined action of these factors governs the distribution observed and the east-west gradients shown in Figure 8.2.

Figure 8.2 shows the horizontal distribution of turbidity in February 1987 at 0 m, 2 m and 3 m with very similar turbidity distribution patterns compared to May 1986, though with slightly higher levels, probably due to dredging or wind effects on the eastern side.

## Seasonal Changes of Turbidity in the North Basin

Figure 8.3 shows the seasonal change of the vertical distributions of turbidity at station P (Figure 8.4). With the exception of June 1983, turbidity in the surface layer has a value of about 1.0 $m^{-1}$ and does not exhibit a major seasonal change. This suggests that the concentration of suspended particles in the surface layer is relatively stable throughout the year in the North Basin. Vertical profiles of turbidity in June, August, and September 1983 and May 1984 in the euphotic layer reflect the vertical distributions of Chl-a concentrations mainly due to autochthonous aquatic biological organisms (Mitamura *et al.*, 1986). More highly turbid waters and frequent turbidity peaks are found in the thermocline region. Turbidity increases are closely related to the postion of the maximum density gradient, which also coincides with a low level of turbulence. The water column below the thermocline is characterized by a low-turbidity layer with a uniform concentration value of about 0.5 $m^{-1}$.

The existence of a turbid benthic nepheloid layer (BNL) close to the bottom, whose character varies greatly during the stratified season, should also be taken into consideration. The presence of the BNL in the North Basin was revealed by observations of concentrated suspended matter (Maeda *et al.*, 1986). While the thickness of the observed BNL does not vary significantly, major changes in concentration are observed in light transmission profiles (Figure 8.5). The properties of the BNL also vary spatially with regards to the summer stratified season (Figure 8.5). Turbidity levels in the bottom layer gradually decrease as the circulation season approaches. The physical mechanisms by which the BNL are maintained have been analysed with data obtained using a turbidity-current meter (Kumagai, 1987) and it has become clear that the suspension mechanisms are closely related to internal seiches. A slight but general increase in turbidity towards the bottom of the North Basin has been verified and this is probably due to the effect of the relatively cold, sediment-laden water from the South Basin flowing into the North Basin as a buoyancy-driven current (Muramoto *et al.*, 1979; Oonishi *et al.*, 1980). However, the relative contribution of each of the internal seiche and density current mechanisms to the creation and maintenance of the BNL is not precisely known.

## Conclusion

The higher turbid water along the east side in the South Basin is the result of a combination of effects, mainly the hydrological conditions, the topography of the South Basin and the effect of human activitity.

The turbidity in the surface layer in the North Basin is relatively stable throughout the year and reflects the vertical profiles of Chl-a concentration and is influenced by the summer stratification. The existence of the benthic nepheloid layer is closely related to internal seiches and bottom boundary currents.

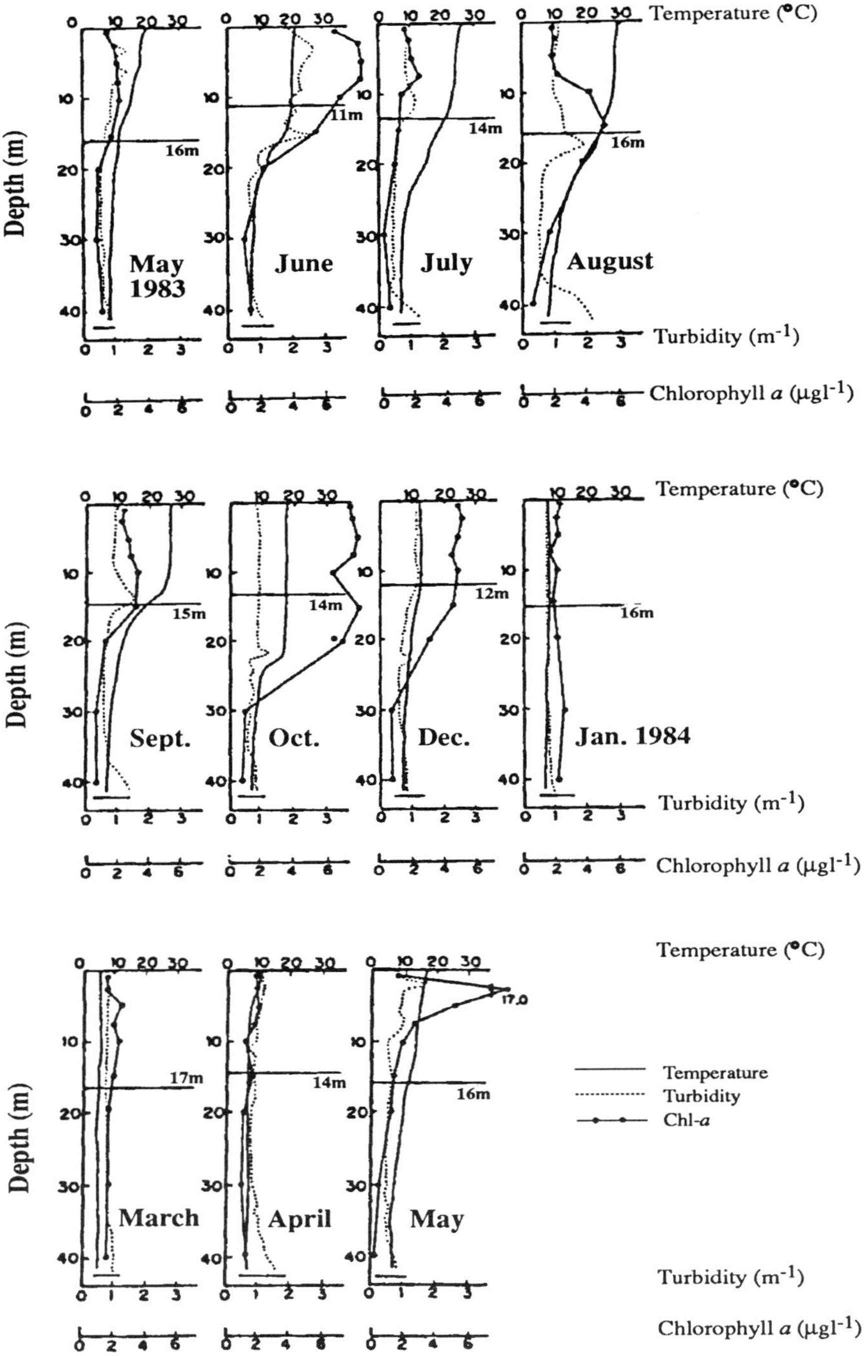


Figure 8.3. Seasonal changes of vertical profiles of beam attenuation coefficient (turbidity) in the North Basin.

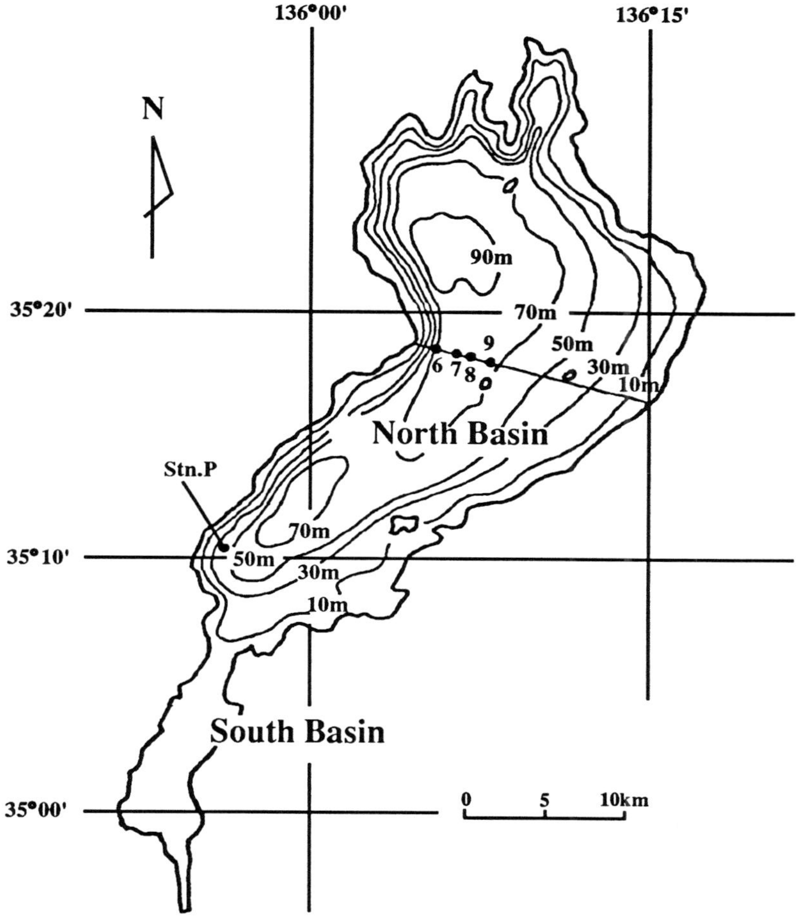


Figure 8.4. Location of stations in the North Basin (Tsuda *et al.*, 1990).

# References


Jerlov, N. G., (1976) *Marine Optics. Elsevier Oceanography* Series 14, Elsevier Scientific Publ. Co. Amsterdam: 1-231.

Kumagai, M., (1987) Bottom current in northern basin of Lake Biwa. *Res. Rep. Lake Biwa Res. Inst.* No. 86-A04: 23-28.

Kumagai, M., H. Maeda, Y. Oonishi, K. Yokoyama and S. Okuda, (1986) Study on effects of low water level by remote sensing. *Res. Rep. Water Resource Res. Center, DPRI, Kyoto Univ.* 7: 39-49.

Kumagai, M., R. Tsuda and T. Kurita, (1990) Analysis of water quality and water temperature in Lake Biwa with remote-sensing technique. *Tech. Rep. Lake Biwa Res. Inst.* No. 89- A06: 12-26.

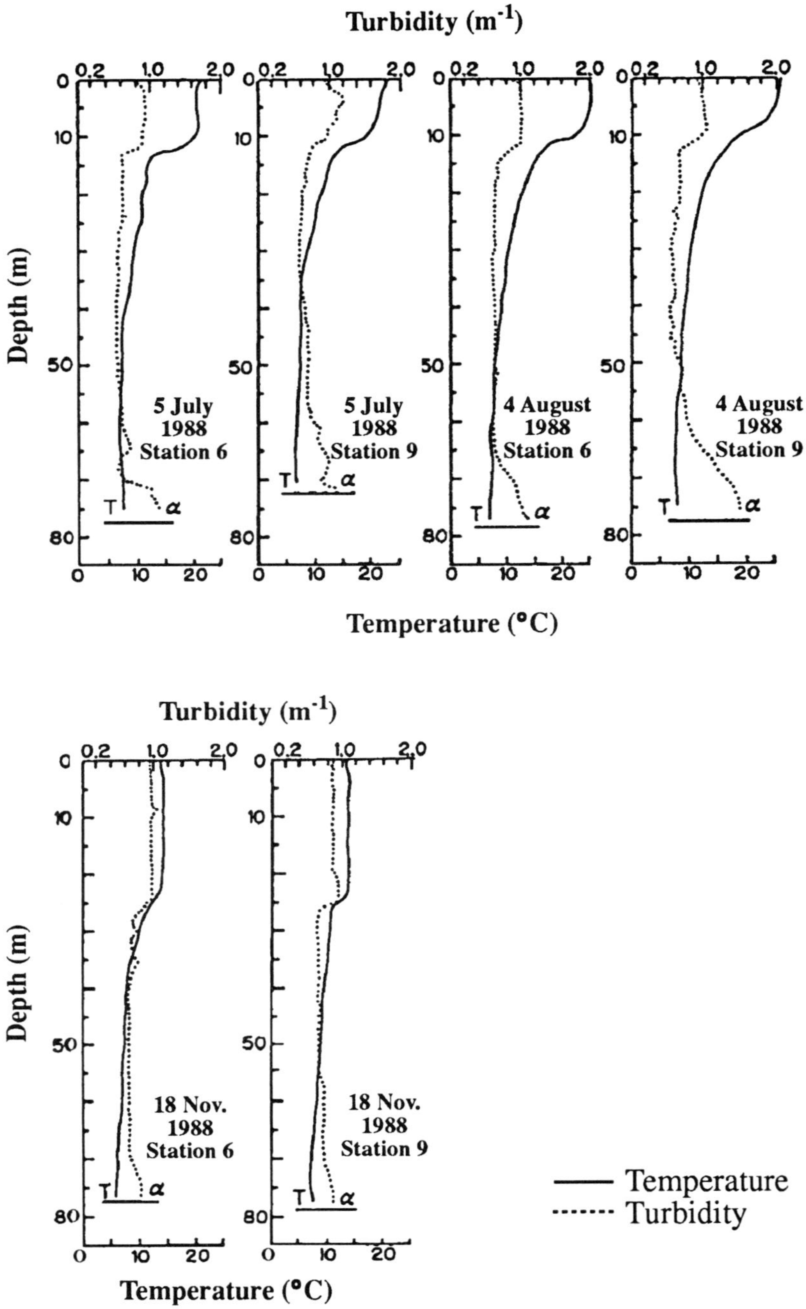


Figure 8.5. Vertical profiles of beam attenuation coefficient in August 1988 in the North Basin (Tsuda *et al.*, 1990).

Muramoto, Y., Y. Oonishi and K. Ookubo, (1979) Heat budget in the South Basin of Lake Biwa - The water exchange between the South and the North Basin of Lake Biwa at Biwa-ko Bridge. *Annuals, Disas. Prev. Res. Inst., Kyoto Univ. No.* 22B-2: 575-588.

Mitamura, O., H. Maeda, T. Kodama and A. Kawai, (1986) The production and decomposition in the North Basin of Lake Biwa. *Res. Rep. Lake Biwa Res. Inst.* No. 86-A05: 11-28.

Maeda, H., O. Mitamura, T. Kodama, M. Eguchi, H. Kitada and A. Kawai, (1986) Sedimentation, decomposition and liberation of substances in the North Basin of Lake Biwa. *Res. Rep. Lake Biwa. Res. Inst.* No.86-A05: 29-74.

Nakanishi, M., T. Narita, O. Mitamura, N. Suzuki and K. Okamoto, (1986) Horizontal distribution and seasonal change of chlorophyll - a concentration in the South Basin of Lake Biwa. *Jpn. J. Limnol.* 47: 155-164.

Oonishi,Y., (1984) *Earth Monthly*, 6: 489-494.

Oonishi, Y., K. Ookubo, A. Harashima and Y. Muramoto, (1980) Coriolis effect on a buoyancy driven current. *Annuals, Disas. Prev. Res. Inst., Kyoto Univ*. No. 23B-2: 351-371.

Okamoto, I., (1964) Current observation of the southern Lake Biwa with a drogue series. *Mem. Fac. Edu., Shiga Univ*. 14: 27-34.

Ookubo, K. and Y. Muramoto, (1986) Current in the Southern Basin of Lake Biwa. Vertical distributions of current velocity and turbidity. *Annuals, Disas. Prev. Res. Inst., Kyoto Univ*. No. 29B-2: 14-25.

Okuda, S., K. Yokoyama, K. Nishi and M. Kumagai, (1986) Study of movement of lake bottom deposit. Part 1. Method and result of observation. *Annual. Disas. Prev. Res. Inst., Kyoto Univ.* No. 29B-1: 341-353.

Smith, R. C and K. S. Baker, (1981) Optical properties of the clearest natural waters (200-800 nm). *Appl. Opt*. 20: 177-184.

Tsuda, R., (1980) Measurements of underwater spectral irradiance in Lake Biwa. *Jpn. J. Limnol.* 41: 57-67.

Tsuda, R., K. Fukagae, M. Kumagai and K. Yokoyama, (1990) Measurements of the nepheloid layer in the North Basin of Lake Biwa. *Tech. Rep. Lake Biwa Res. Inst*., No. 89-06 : 1-11.

Tsuda, R. and M. Nakanishi, (1990) Spectral distribution characteristics in downwelling irradiance and optical properties within the euphotic layer in the North Basin of Lake Biwa. *Arch. Hydrobiol*. 119:279-292.

# 9

# Dynamics of the Turbid Benthic Boundary Layer

M. Kumagai, R. Tsuda and K. Fukagae

## Introduction

In a recent review of studies on benthic boundary layers (BBL), Bedford and Abdelrhman (1987) described three types of BBL observed in lakes: steady boundary layers, unsteady boundary layers and suspended sediment boundary layers.

The first type, defined as a steady boundary layer, contains diffusive and conductive layers, viscous sublayers, turbulent logarithmic layers, and Ekman layers at the bottom. For example, Sailor and Miller (1988) measured profiles of current speed at the bottom of Lake Michigan and found evidence of logarithmic current profiles as well as Ekman veering of the velocity vectors.

The second type, defined as an unsteady boundary layer, is formed by three processes: by long period waves including seiches, by surface short period gravity waves, and by internal waves during the stratified season. This type of boundary layer is described by Grant and Madsen (1986) as a wave-current boundary layer and Kumagai (1986) applied this model in his studies on sediment transport in Lake Biwa.

The third type, defined as the suspended sediment boundary layer, occurs when the dynamics are influenced by both the background velocity field and the presence of suspended sediment. This boundary layer can exist at various scales in time and space. For example, Kumagai (1988) examined the resuspension of bottom sediment caused by wind waves in the shallow coastal area (depth 2 m) of Lake Biwa and deduced a bottom

Physical Processes in a Large Lake: Lake Biwa, Japan
Coastal and Estuarine Studies, Volume 48, Pages 87–99

turbid layer thickness of about 0.02 cm. On the other hand, Sandilands and Mudroch (1983) measured a BBL called the nepheloid layer in Lake Ontario and established a thickness of several tens of meters. Generally speaking, the thickness of the suspended sediment boundary layer seems to be determined by physical parameters such as the period of the dominant waves, the friction velocity, the eddy diffusivity and the settling velocity of the sediment.

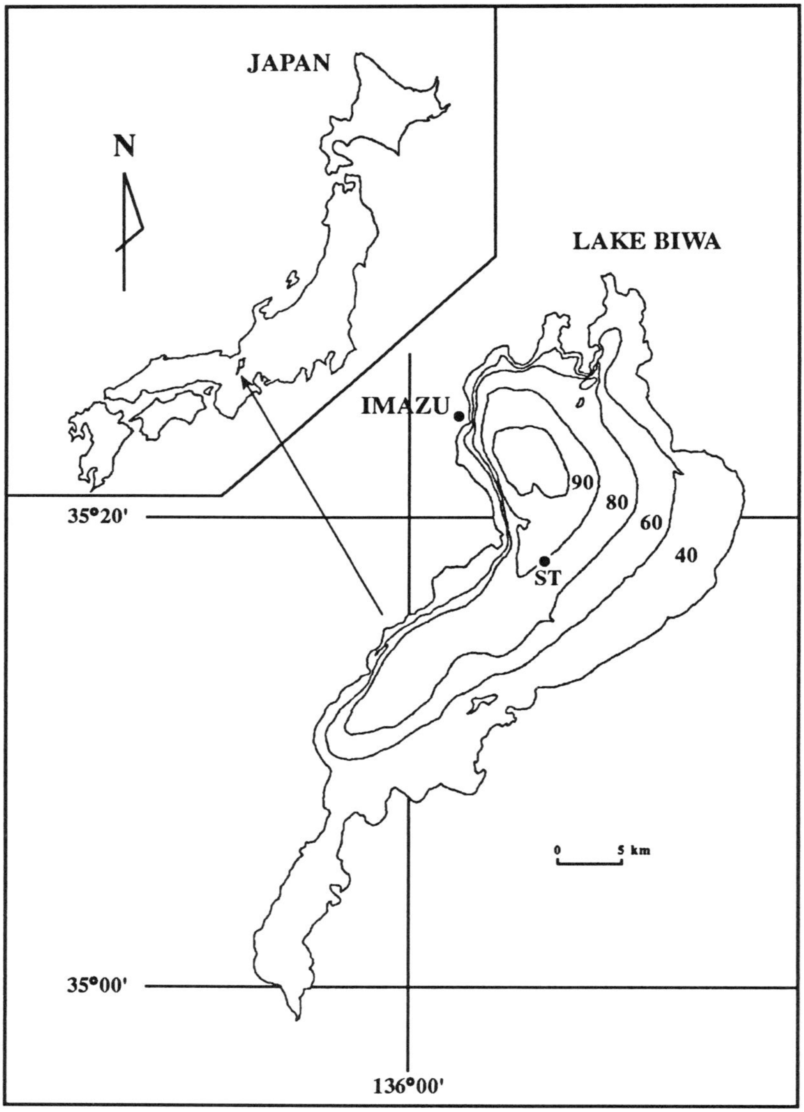


Figure 9.1. Bathymetry of Lake Biwa and location of the observation station.

In order to clarify the mechanisms that sustain the turbid benthic boundary layer in Lake Biwa, field observations were conducted with 9 current meters (Alec Magnetic Current Meter: ACM4M), capable of measuring three dimensional current velocities, turbidity and water temperature, and 1 TCDO meter (Sea Bird; Temperature, Conductivity and Dissolved Oxygen : SBE16). The accuracies of the three velocity components of the ACM4M are 0.6 $cms^{-1}$ for horizontal velocities and 0.3 $cms^{-1}$ for the vertical component, with velocity thresholds of 0.1 $cms^{-1}$ in each direction. The ACM4M measures turbidity with an accuracy of 0.7 mg $L^{-1}$, and water temperature with 0.03 °C. Similarly, the accuracy of each sensor of SBE16 is as follows: water temperature 0.004 °C, conductivity 1μS/cm, and dissolved oxygen 0.1 ml $L^{-1}$.

# Observations

The observations were performed at station ST indicated in Figure 9.1 between November 15 and 27, 1988. The depth of the station was 80 m and the instruments were moored as shown in Figure 9.2. The current meters were placed at heights of 0.9, 3.2,

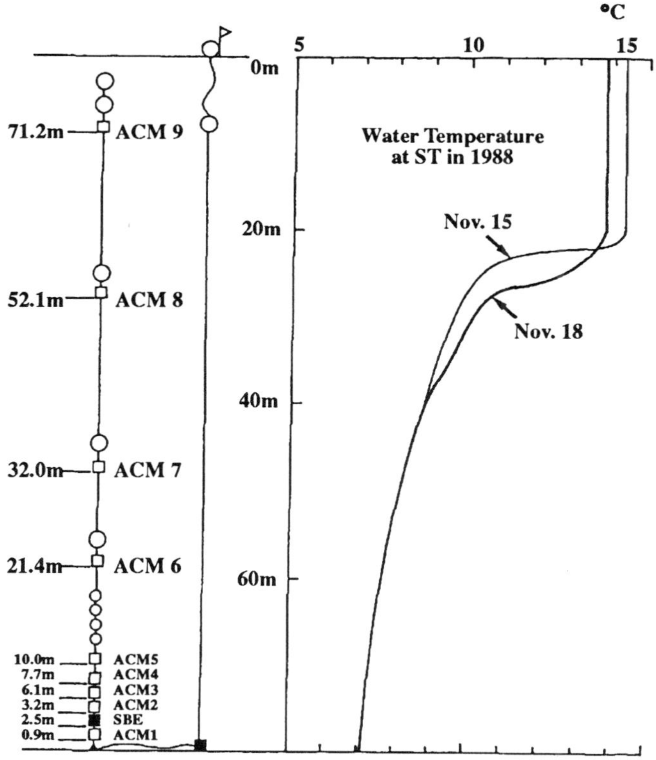


Figure 9.2. Configuration of the instrument moorings.

6.1, 7.7, 10.0, 21.4, 32.0, 52.1, and 71.2 m above the lake bottom, while the TCDO meter was placed at a height of 2.5 m. All data were recorded in burst-mode sampling at intervals of 20-minutes and stored in IC memories. The current meters sampled 30 data points over 30 second intervals while the TCDO recorded 1 sample at each interval. Water temperature profiles measured on November 15 and 18 are shown in Figure 9.2 and indicate a deepening of the thermocline between the two days.

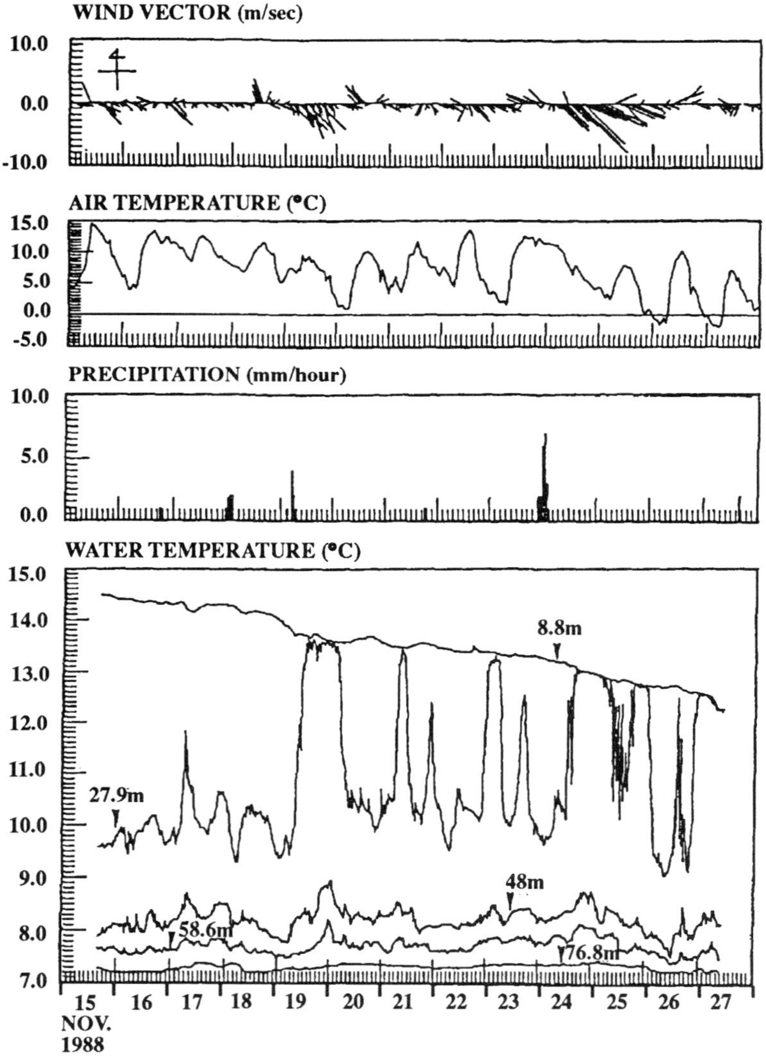


Figure 9.3. Temporal change of wind vectors, air temperature, precipitation, and water temperature. Meteorological data was gathered at Imazu, and water temperature was measured at the station in Lake Biwa.

Meteorological data over the study period are shown in Figure 9.3. The data were measured at Imazu located on the northwest coast of Lake Biwa, and water temperatures were obtained from 5 ACMs (ACM2, ACM6, ACM7, ACM8, and ACM9) at ST. As seen in Figure 9.3, there was a strong northwest wind of over 5 $ms^{-1}$ blowing on two occasions during the period of observation, on November 19 and 24. A rapid decrease in air temperature followed these strong winds in the early mornings of November 20 and 25. In addition, heavy rain fell from before midnight on November 23 until early morning, with a total precipitation of 22 mm for the period.

The last panel in Figure 9.3 shows temporal variation of water temperature at depth of 27.9 m. A sudden rise in water temperature of more than 4 °C at 27.9 m(ACM8) occurred on November 19 immediately following the strong wind event, which was obviously caused by the excitation of internal waves. As seen in Figure 9.3, the water temperature at 27.9 m oscillates with a period of almost 2 days, and is most likely an internal Kelvin wave (see also Figure 9.7). It is interesting that the temporal change in water temperature at depths of 48.0 m and 58.6 m show different dominant waves with much shorter periods of 2 or 3 hours due to excitation by smaller scale internal waves. The short wave period of 2 - 3 hours is related to the Brunt-Väisälä frequency, while the 60 hour wave period corresponds to internal Kelvin waves. The water temperature at a depth of 76.8 m did not seem to vary significantly. Nevertheless, this does not mean that waves were absent near the bottom since the vertical temperature gradients were very weak at depth (Figure 9.2).

In Figures 9.4 and 9.5, the N-S and E-W components of horizontal current speeds are depicted. ACM9 measured current in the epilimnion above the thermocline, ACM8 was placed in the thermocline, and all others recorded current speeds over time in the hypolimnion below the thermocline with the distances measured from the bottom. For the five current meters (ACM1 to ACM5) within 10 m of the bottom, both the N-S and E-W measured components sometimes show strong currents with a speed of around 10 $cms^{-1}$ in each component near the bottom. Two kinds of oscillation, one with a long period (about 2 days) and the other with a short period (2-3 hours) are also evident in these time series near the bottom. The short period oscillations are particularly clear in the N-S component of current speed measured by ACM3 and ACM4 between November 15 and 18. In the thermocline horizontal velocities were relatively low, as seen in the data obtained from ACM8, while above the thermocline (ACM9) currents were 180° out of phase with currents below the thermocline. From November 19 to 27, the rotation of the current vector above the thermocline was counter-clockwise, while that below the thermocline was clockwise (see Kumagai, 1986), consistent with currents expected in connection with the phenomenon of internal Kelvin waves (e.g. Csanady, 1984).

## Turbid Benthic Boundary Layer

Due to the locally high levels of turbidity, the dynamics of the layer within 10 m of the bottom differ from those higher in the water column. This layer is called the turbid benthic boundary layer (TBBL), and in Lake Biwa usually forms in May and disappears in December, coinciding with the period over which the thermocline is generated. A similar TBBL was observed in Lake Ontario by Sandilands and Mudroch (1983). Recently, as part of the High Energy Benthic Boundary Layer Experiment (HEBBLE), bottom turbidity was measured over time in 4800 m of water on the Nova Scotian continental rise, and found many "suspended sediment storms" induced by intermittent turbulence near the bottom.

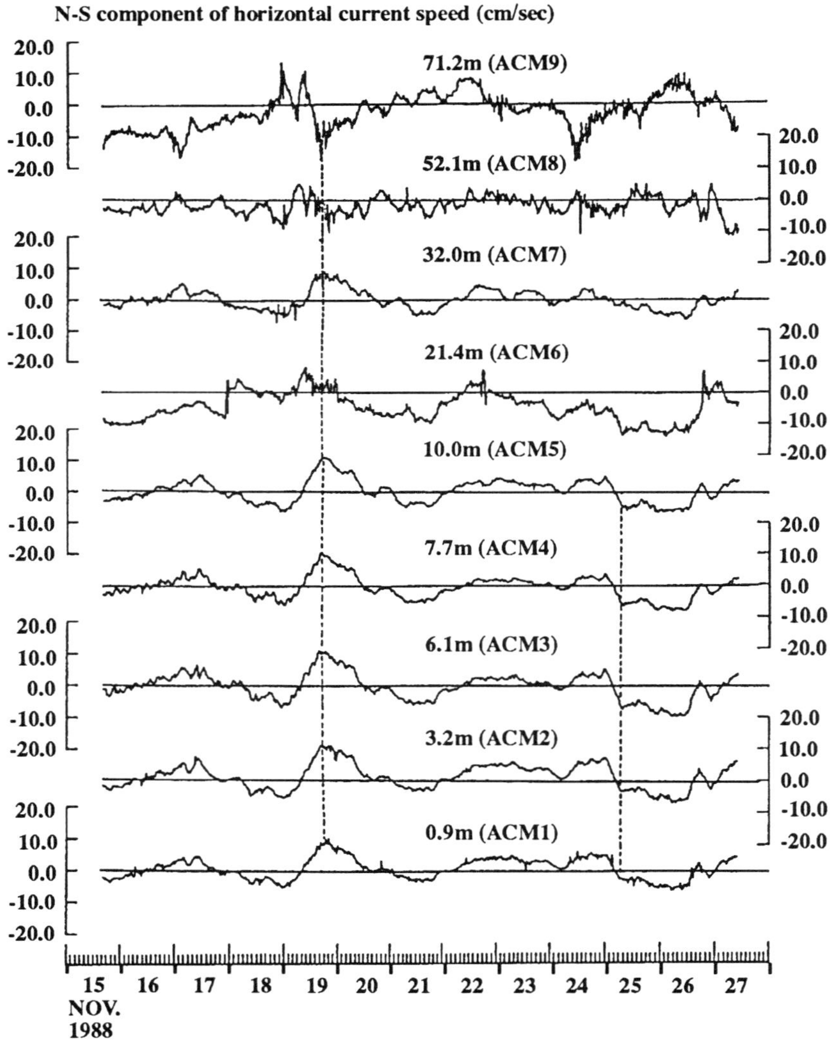


Figure 9.4. Temporal changes of the N-S component of horizontal current speed at 71.2, 52.1, 32.0, 21.4, 10.0, 7.7, 6.1, 3.2, and 0.9 m above the bottom measured between November 15 and 27, 1988.

Two unresolved questions remain regarding the TBBL : the first is the source of the turbidity, and the second is the mechanism maintaining the turbidity in the TBBL. It is difficult to clarify precisely, whether the TBBL is stable or unstable. If it is stable, the turbidity levels must be maintained by a steady or effectively steady source of turbidity. Since it has been established that the TBBL exists during the stratified season, it seems to be stable and turbidity seems to be replenished frequently by some mechanism.

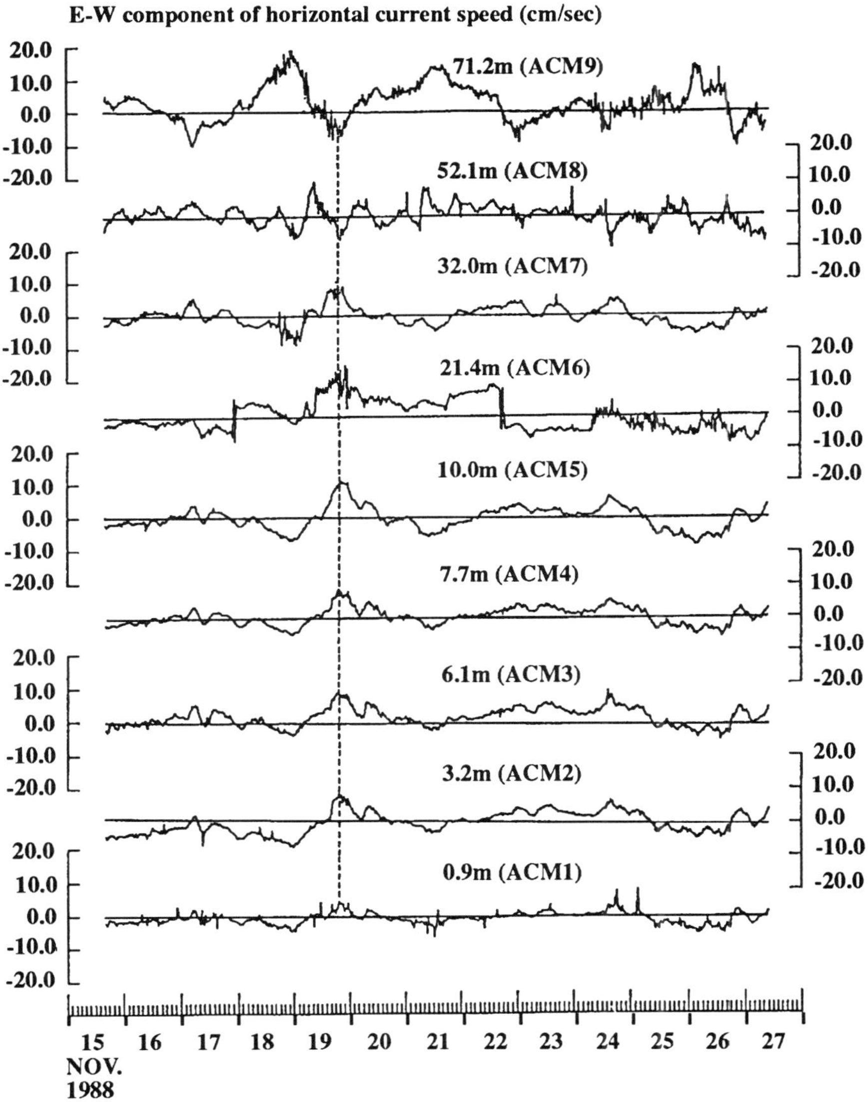


Figure 9.5. Temporal changes of the E-W component of horizontal current speed.

Kawana and Tanimoto (1981) pointed out that the tidal current induces a TBBL in the Seto Inland Sea, but the notion that a persistent strong current may induce the TBBL of Lake Biwa is not supported by the data provided in Figures 9.4, 9.5, and 9.6. For example, a strong northeast current of about 10 $cms^{-1}$ occurred near the bottom on November 19, but it resulted in a decrease of turbidity (Figure 9.6). Conversely when turbidity in the TBBL was high, as from November 15 to 18 for example, currents in the TBBL were relatively low. If turbidity is supplied from the bottom by vertical diffusion only, a turbidity gradient should have been detected by the 5 ACMs (ACM1 to ACM5)

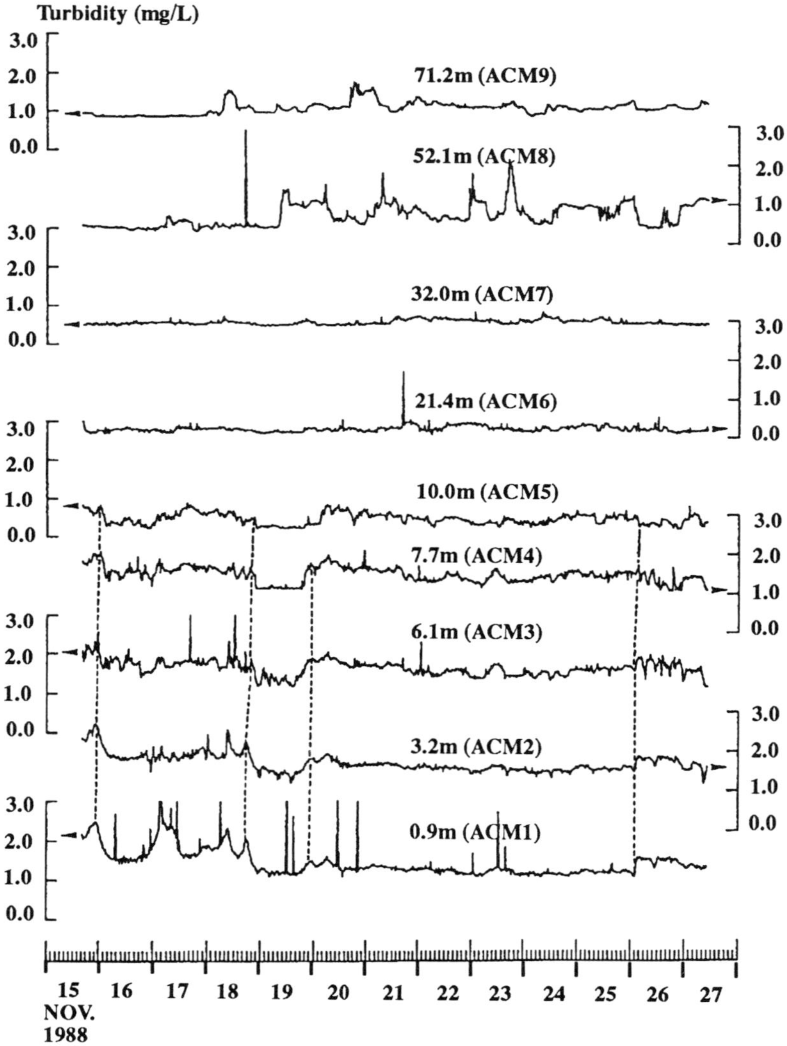


Figure 9.6. Temporal changes of turbidity.

but this could not be verified, thus implying the occurrence of an intermittent resuspension of bottom sediment and many spikes in turbidity seen at ACM1, ACM2, and ACM3 support this assumption. More detailed observations are required, however, before definite conclusions concerning the source of turbidity can be attained.

Once turbidity is present in the boundary layer, little energy is apparently needed to maintain the net sediment load of a TBBL throughout the summer since deposition rates are so low (Halfman and Johnson, 1989). Events such as wind-driven bottom currents, surface waves, internal waves, and even fish may well supply enough energy to maintain the sediment in suspension. In order to determine what maintains a TBBL, the MEM (Maximum Entropy Method) spectrum of water temperature at ACM8 and the MEM spectra of turbidity of ACM1-ACM5 were calculated. The results are shown in Figure 9.7, with the top curve indicating the MEM spectrum of water temperature, and the others showing the MEM spectra of turbidity. The period of 56.4 hours in the spectra of water temperature corresponds to the internal Kelvin wave, and the other short periods may be due to internal seiches or small scale internal waves. It should also be noted that only high frequency waves of a period of several hours exist in turbidity changes near the bottom. As seen in Figure 9.4 and Figure 9.5, the current changes near the bottom were found to have a period of the internal Kelvin wave.

Figure 9.8 presents the vertical profiles interpolated from the 9 point measurements in the vertical of both, (a) the N-S and (b) the E-W components of horizontal current speed, (c) the vertical current speed, (d) the water temperature, (e) the Brunt-Vaisala frequency, and (f) the turbidity measured from November 15 to 18. From Figure 9.8 (f) TBBL can be clearly recognized below the depth of 65 m. The maintaining mechanism of the TBBL is still unknown, but these data show two possibilities, one is the existence of a strong shear flow in BBL which may induce the convergence of momentum near the upper boundary of BBL (see Figure 9.8 (a) and (b)). Another is, as shown by Kanari (1984), upward vertical velocity from internal waves may occur by Brunt-Vaisala frequency near the bottom. Figure 9.8 (c) and (e) seem to support this idea. As can be recognized in Figure 9.8 (c), there are three nodal points of vertical current speed at depths of about 30 m and 65 m, which may correspond to the vertical velocity of the fourth mode internal wave (Kanari, 1984). Figure 9.8 (d) shows a small thermocline above the BBL and water temperature is almost constant in the BBL.

## Dissolved oxygen in the TBBL

Recently, the concentration of dissolved oxygen has been decreasing near the bottom of Lake Biwa. In order to examine the relationship between TBBL and dissolved oxygen, a TCDO meter was placed 2.5 m above the bottom. Temporal changes of dissolved oxygen (TCDO) and turbidity (ACM2) are shown in Figure 9.9. It is interesting that the changes of turbidity are strongly negatively correlated with changes of dissolved oxygen, i.e., turbidity is high when dissolved oxygen is low, and vice versa. Previously, dissolved oxygen was not considered to vary over a period as short as one day.

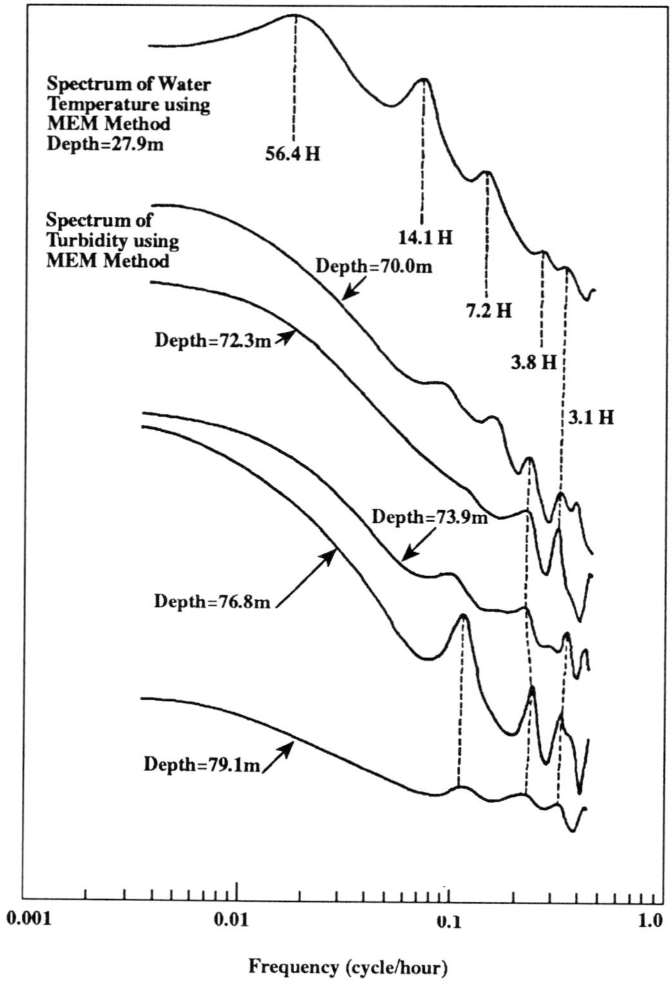


Figure 9.7. MEM spectra of water temperature at a depth of 27.0 m and turbidity at depths of 70.0, 72.3, 73.9, 76.8, and 79.1 m.

A plot of dissolved oxygen against turbidity is shown in Figure 9.10, clearly indicating the presence of an hysteresis effect. In particular, as turbidity increases, the decline in dissolved oxygen starts slowly and then accelerates quickly. In the same way, as turbidity decreases, the recovery of dissolved oxygen starts slowly and then suddenly accelerates. This phenomenon may be due to the fact that the TBBL is moving. For example, as a water mass with high turbidity and low dissolved oxygen approaches the station, dissolved oxygen concentration is higher at the head of the water mass than at the tail. This is due to the fact that the head is always in contact with water containing rich dissolved oxygen, while the tail of the water mass follows water with low dissolved

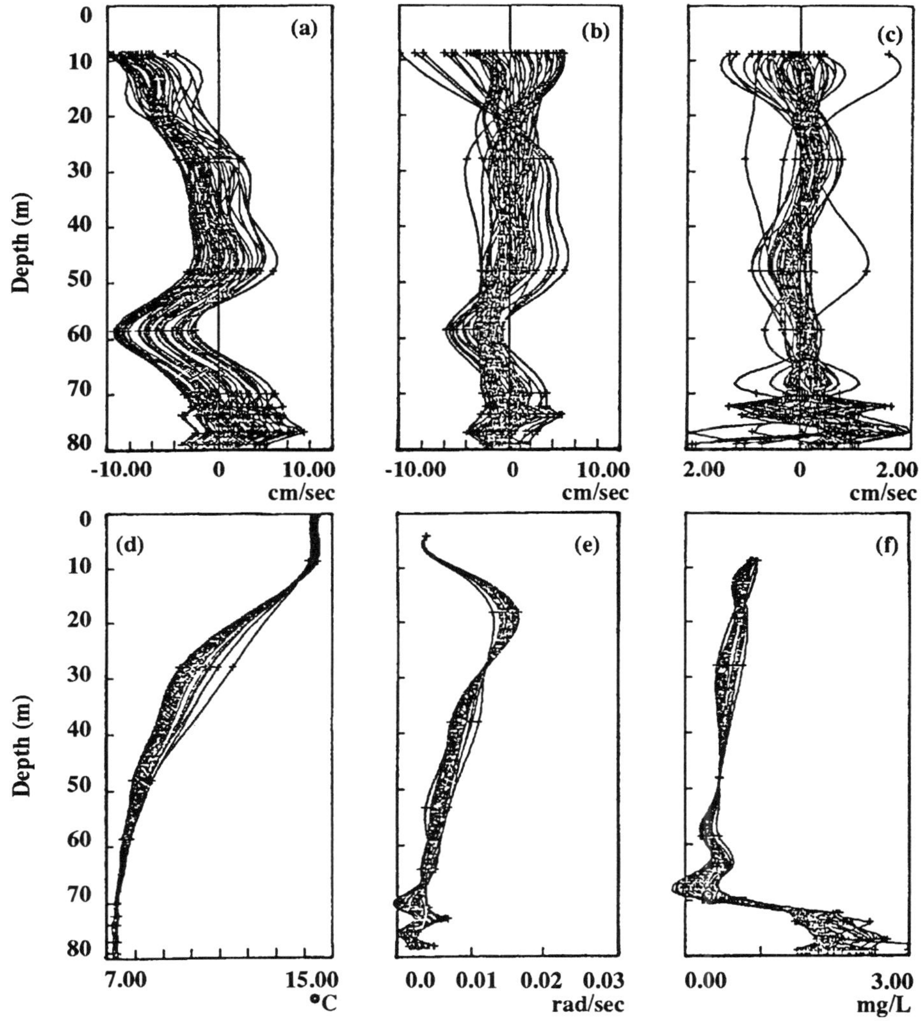


Figure 9.8 . Vertical profiles of (a) N-S component of current speed, (b) E-W component of current speed, (c) vertical current speed, (d) water temperature, (e) Brunt-Vaisala frequency, and (f) turbidity measured from November 15-18, 1988.

oxygen. Thus, when the passing of the water mass at a fixed point is measured, we first sense higher turbidity and richer dissolved oxygen followed by higher turbidity and poorer dissolved oxygen. Therefore, near the bottom during calm weather, the TBBL seems to move with a period of about one day (Figure 9.9).

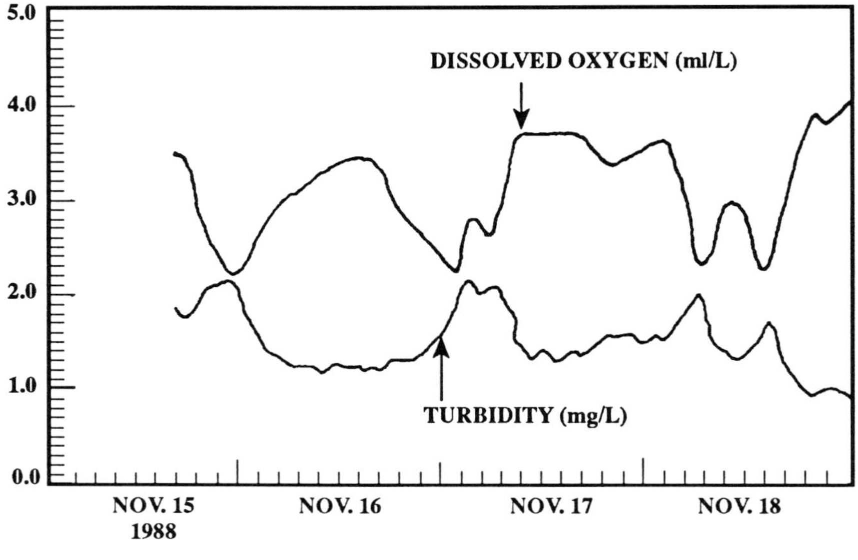


Figure 9.9. Temporal change of dissolved oxygen at 2.5 m above the bottom and of turbidity at 3.2 m above the bottom.

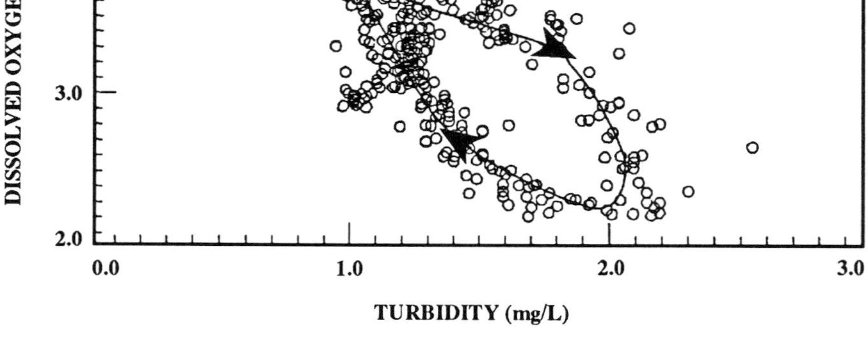


Figure 9.10. Relationship between dissolved oxygen and turbidity in the TBBL.

# Conclusion

Observations by current and TCDO meters moored in Lake Biwa in 1988 showed the presence of internal waves with periods of 56.4, 14.1, 7.2, 3.8, and 3.1 hours. The longest internal wave is the internal Kelvin wave, and this produced a counter-clockwise current in the epilimnion and clockwise current in the hypolimnion at the measurement site.

A turbid benthic boundary layer (TBBL) of thickness about 10 m was found near the bottom. The turbidity in the TBBL is not supplied from the bottom by shear stress, but is supported by an intermittent resuspension of bottom sediment with a patchy structure moving with a period of several hours. A negative correlation between turbidity and dissolved oxygen was found to exist in the TBBL, and suggests that the TBBL fluctuates with a period of about one day.

# References

Bedford, K. W. and M. Abdelrhman, (1987) Analytical and experimental studies of the benthic boundary layer and their applicability to near-bottom transport in Lake Erie. *J. Great Lakes Res.* 13(4): 628-648.

Csanady, G. T., (1984) Milestones of research on the physical limnology of the Great Lakes. *J. Great Lakes Res.* 10(2): 114-125.

Grant, D. W. and S. Madsen, (1986) The continental-shelf bottom boundary layer. *Ann. Rev. Fluid. Mech.*, 18: 265-305.

Ivey, G. N. and R. I. Nokes, (1989) Vertical mixing due to the breaking of critical internal waves on sloping boundaries. *J. Fluid Mech.* 204: 479-500.

Halfman, B. M. and T. C. Johnson, (1989) Surface and benthic nepheloid layers in the western arm of Lake Superior, 1983. *J. Great Lakes Res.* 15(1): 15-25.

Kanari, S. (1984) Internal waves and seiches. *Lake Biwa* edited by S. Horie. Dr. W. Junic Publishers, 185-235.

Kawana, K. and T. Tanimoto, (1981) Temporal variation of resuspended matter for one tidal cycle in the Hiuchi-Nada area. *J. Oceanogr. Soc. Jap.*, 37: 173-178.

Kumagai, M., (1986) Bottom current in the Northern Basin of Lake Biwa. *Res. Rep. Lake Biwa Res. Inst.* No.86-A04: 23-28.

Kumagai, M., (1988) Predictive model for resuspension and deposition of bottom sediment in a lake. *Jap. J. Limnol.*, 49(3): 185-200.

Lynch, J. F., T. F. Gross, B. H Brumley and R. A. Filyo, (1988) Sediment concentration profiling in HEBBLE using a one megahertz acoustic backscatter system. *Deep Sea Res.* (in press).

Sailor, J. H. and G. S. Miller (1988) Observation of Ekman veering at the bottom of Lake Michigan. *J. Great Lake Res.* 14(1): 94-100.

Sandilands, R. G. and A. Mudroch, (1983) Nepheloid layer in Lake Ontario. *J. Great Lakes Res.* 9(2): 190-200.

# 10

# Distribution and Mixing of the River Seri Water

I. Okamoto, S. Endoh and M. Kumagai

## Introduction

Dispersion processes associated with discharged water from power plants have often been discussed because of their heavy impact on coastal environments. When a river flows into the sea, the river water always spreads over the sea surface, but when a river flows into a fresh water body, a variety of inflow configurations are possible depending upon the different densities of the river water compared to that of the lake water (Lindell, 1980). In the cold winter months when river water is colder than lake water, river water flows beneath the lake water as a density current (Anderson and Pritchard, 1951; Kao, 1977). Conversely, in the warm summer months the river water is warmer than lake water and the river water spreads out on the surface of the lake. After flowing into the lake basin, the river water is transported by lake currents or mixes with the lake water. Coastal mixing as well as the lake-river interaction has been the subject of intensive research by Churchill (1985), Fischer *et al.* (1980), Han and Niedrauer (1981), and Carmack *et al.* (1979). Likewise, Ishitobi *et al.* (1989) discussed the intrusion of sea water into a lake basin and the effect of tidal mixing processes in his recent paper.

In the following, the distribution and dispersion of the River Seri (catchment area: 65 $km^2$) as it flows into Lake Biwa, will be discussed. The water from the River Seri has much higher electrical conductivity than the Lake Biwa water due to the presence of a calcarious soil zone in the catchment. This conductivity difference was therefore used as a tracer to follow the inflowing river water as it entered the lake.

Physical Processes in a Large Lake: Lake Biwa, Japan
Coastal and Estuarine Studies, Volume 48, Pages 101–118

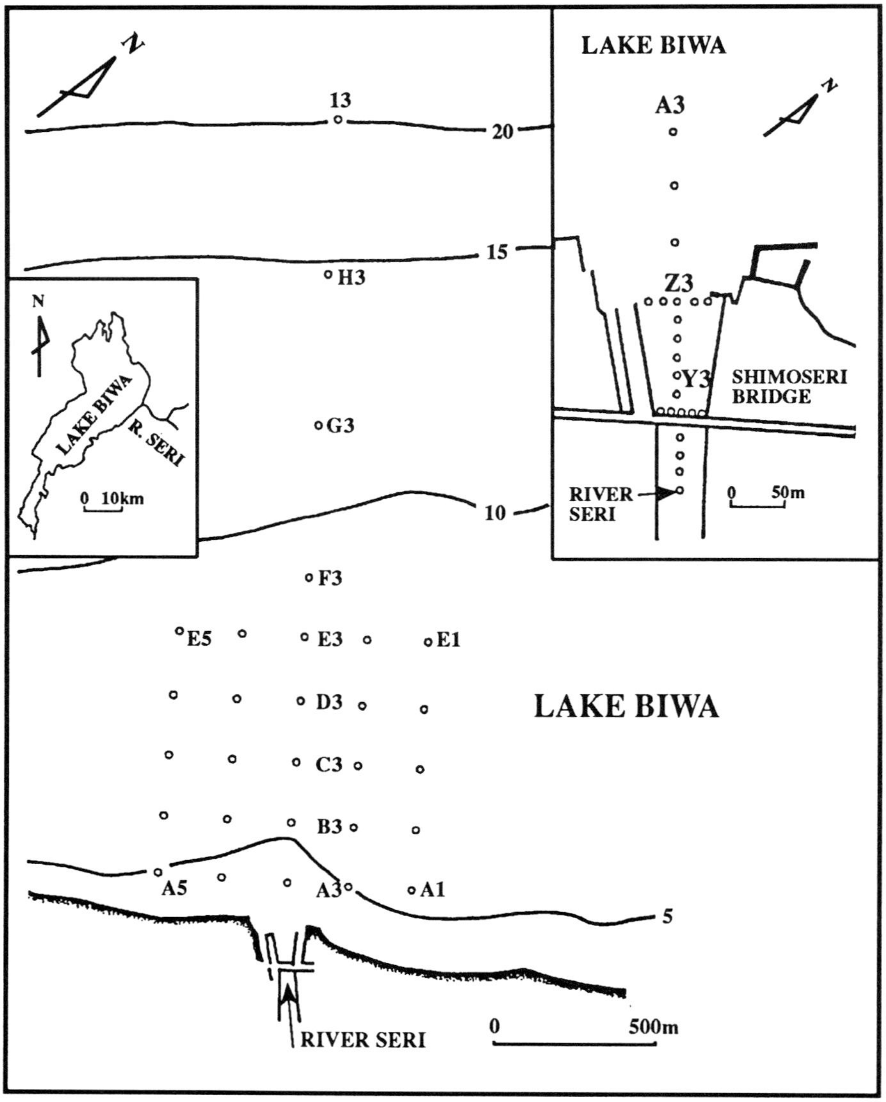


Figure 10.1. Locations of observation stations near the mouth of the River Seri. Depth contours are in meters.

## Observations

Water temperature, electrical conductivity (at a reference of 18°C) and discharge (water level) have been continuously monitored at fixed stations along the River Seri for some time. In addition, measurements have been made from two boats, one surveying temperature, conductivity and current velocities in the region near the river mouth, while the other was used to collect data along the length of the river. In view of the highly variable nature of the flow, the observations were usually repeated twice a day during the sounding period and observations were carried out not only during an ordinary discharge period, but also in the flood season.

About ten self-recording instruments were set up at various stations arranged in a relatively small area near the river mouth as shown in Figure 10.1. These instruments recorded temperature, conductivity, turbidity and current velocity. In order to maintain accurate positioning of observation stations and boats, marker buoys were employed and the position of the boat was periodically fixed by using theodolites from the land.

## Front Formation and Behavior Near the River Mouth

Fronts usually formed at the boundary between the river and the lake water and they were modified or influenced by local lake-land breezes. Figure 10.2 shows four sets of conductivity and temperature distributions for a vertical section along the river over a two day period. A well-defined front existed at the boundary between the river water (high conductivity) and the lake water (low conductivity) on the two successive days. The front was highly variable in structure although two basic frontal types were observed. In the first, river water overflowed the lake water (Figures 10.2 a1, b1 and a3, b3), and in the second, river water underflowed the lake water (Figure 10.2 a2, b2 and a4,b4); in the morning there was an overflow and in the afternoon an underflow.

Both patterns are produced under the influence of lake-land breezes. In the first case, land breezes push the river water offshore and deep lake water intrudes into the river bottom. In the second case, lake breezes arrest the propagation of the front at the surface, and the river water moves offshore under the lake water. These two patterns of front formation shown in Figure 10.2 are produced under the influence of the regular daily lake-land breezes typical of fine days. There appears to be considerable mixing at this front and as a consequence the river water is considerably diluted before leaving the river mouth.

The dynamic behavior of this front can be seen in the time series records shown in Figure 10.3 from the near-bottom region at the anchor stations. At midnight on June 18, 1989, the temperature dropped almost simultaneously at the anchor stations (E-2, E-3 and E-4) with a concurrent sudden rise in conductivity as shown by the arrows in Figure 10.3. At Station G-3 a similar effect occurred about two hours earlier. All sensors were placed 25 cm above the lake bottom, with the exception of the second tracer at Station E-3 ("near bottom") positioned at 100 cm above the bottom. Progressive vector diagrams established from these data, show that the front moved offshore with a velocity of about 5.0 $cms^{-1}$ (Figure 10.4). The sudden change in temperature and conductivity at Station G-3 indicated another front also advancing offshore.

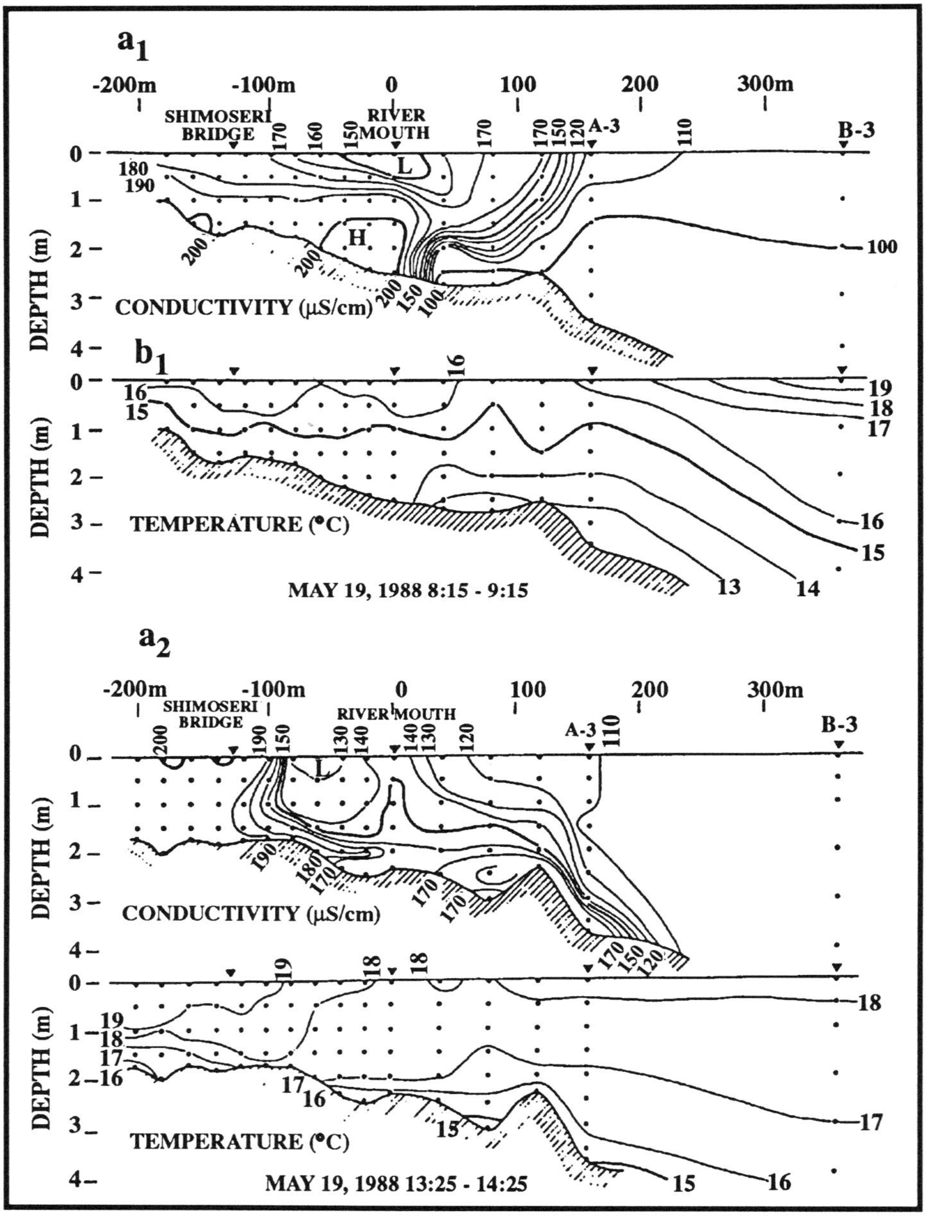


Figure 10.2. Drastic changes in cross-sectional distributions of conductivity ($a_1$-$a_4$) and temperature ($b_1$-$b_4$) near the river mouth on two successive days.

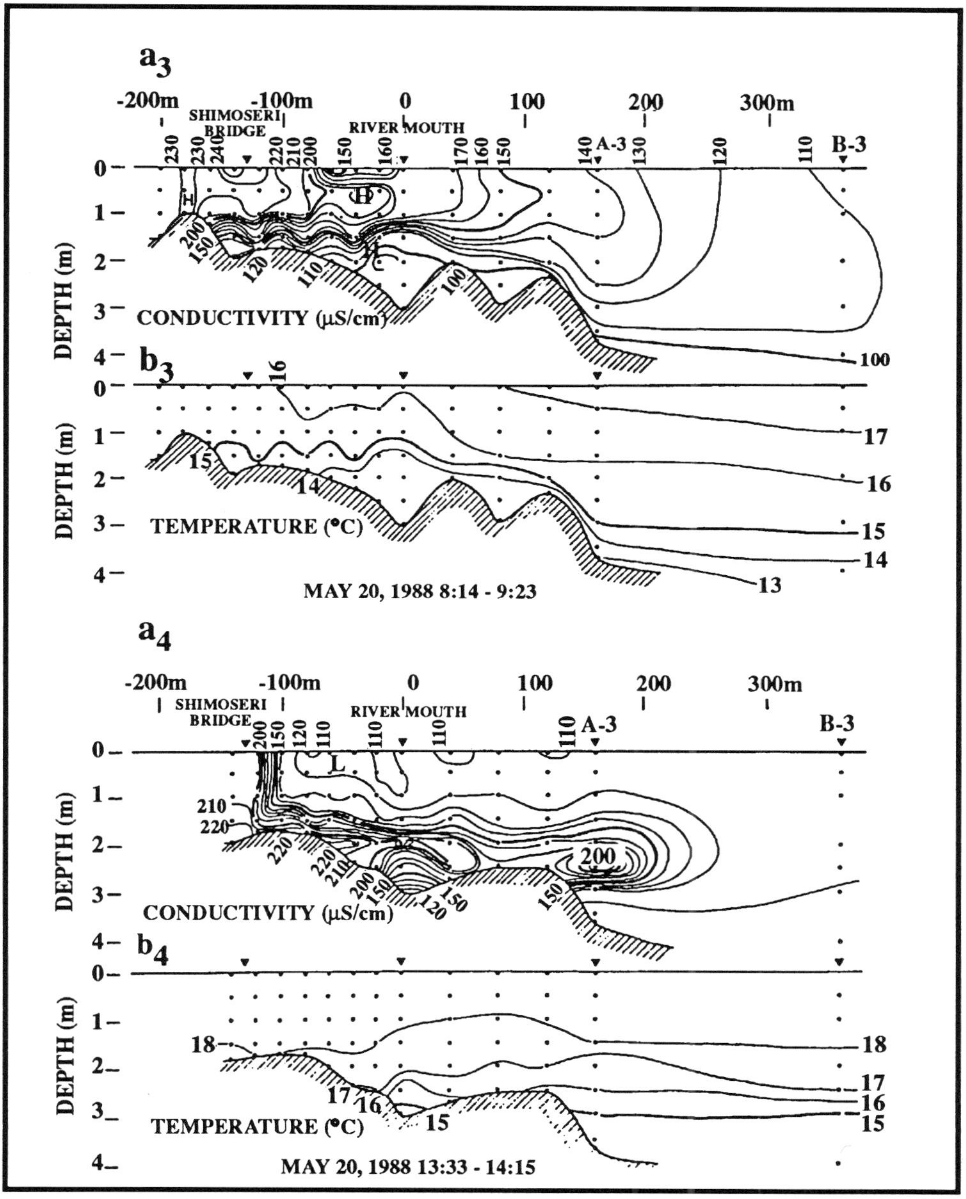


Figure 10.2. (Continued)

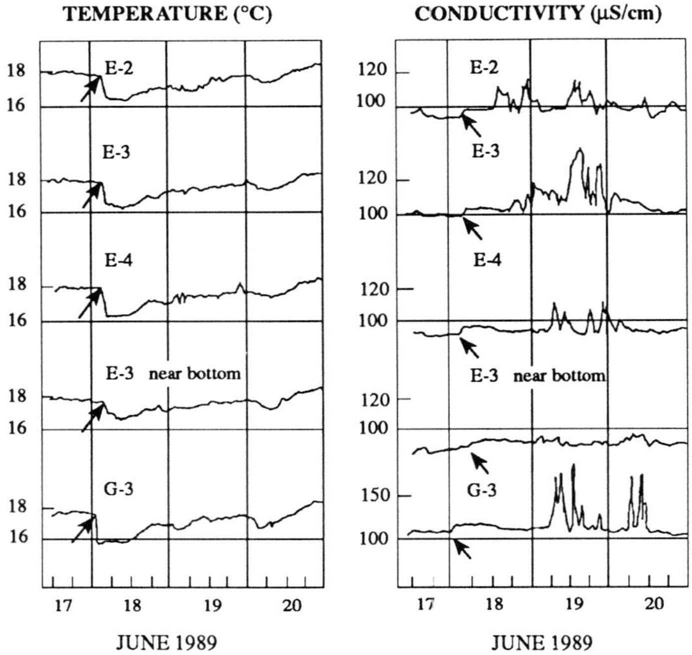


Figure 10.3. Sudden changes in conductivity and temperature in the bottom layer at midnight, June 18, 1989.

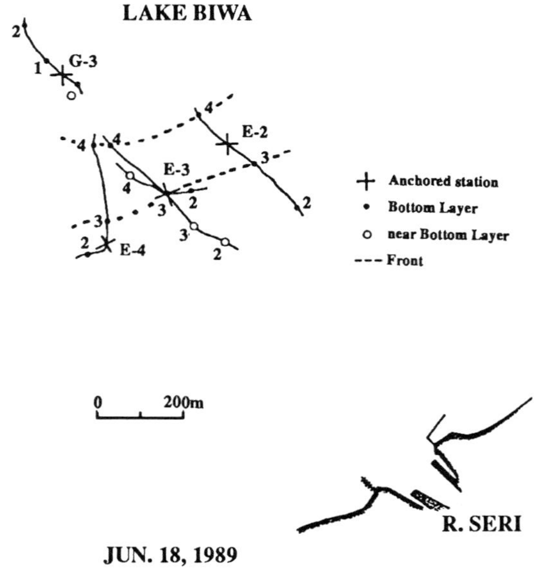


Figure 10.4. Progressive vector diagrams immediately before and after the appearance of a front advancing offshore, June 18, 1989 (cf. Figure 10.3).

## Underflow and Overflow of River Water into the Lake

The inflow pattern is variable. Under the conditions when there is little or no difference in temperature between river and lake water, the inflowing pattern is influenced by local wind as already shown in the previous section. Under calm or weak wind conditions the inflowing river enters the lake at the surface (overflow) or at the bottom (underflow) depending on the temperature (density) difference between river and lake. Particularly after heavy rains, the river water is usually much colder than normal in the stratified season producing an underflow. In this case, the cold high conductivity river water lies on the bottom with very little vertical mixing and often penetrates into the thermocline at its level of neutral buoyancy when flowing into deeper waters of the lake (Figure 10.5).

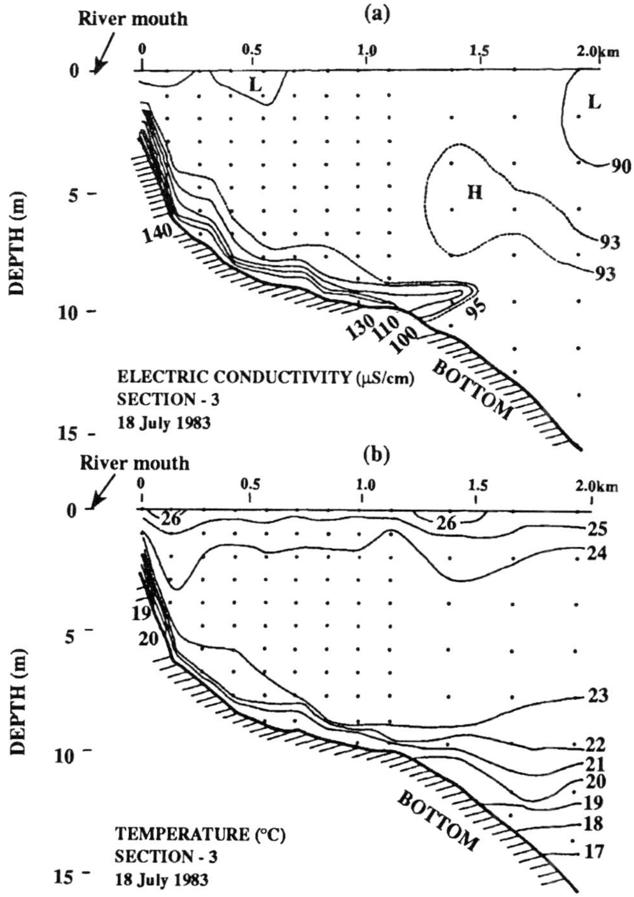


Figure 10.5. An example of the cross-sectional distribution of river water beneath lake water. A progressive head of river water was found to penetrate into the thermocline.

By contrast, warm river water with a high conductivity rate (>100 μS/cm) spreads over the lake surface as shown in Figure 10.6. The overflow is strongly influenced by wind waves and convective cooling at night, with both effects tending to promote the mixing of river and lake water.

An example of a time series of temperatures in the river water and lake water is presented in Figure 10.7. The amplitude of the daily fluctuations in the river water temperature is larger than that of the lake water. Even in the warm season the river water is often colder than the lake water at night, so that both underflow and overflow can occur each day. With the exception of abnormal declines of the lake water temperature caused by upwelling of deep water, such as that observed on September 25, 1982, the lake water is warmer than the river water in the winter months and thus leads primarily to the occurrence of underflow patterns.

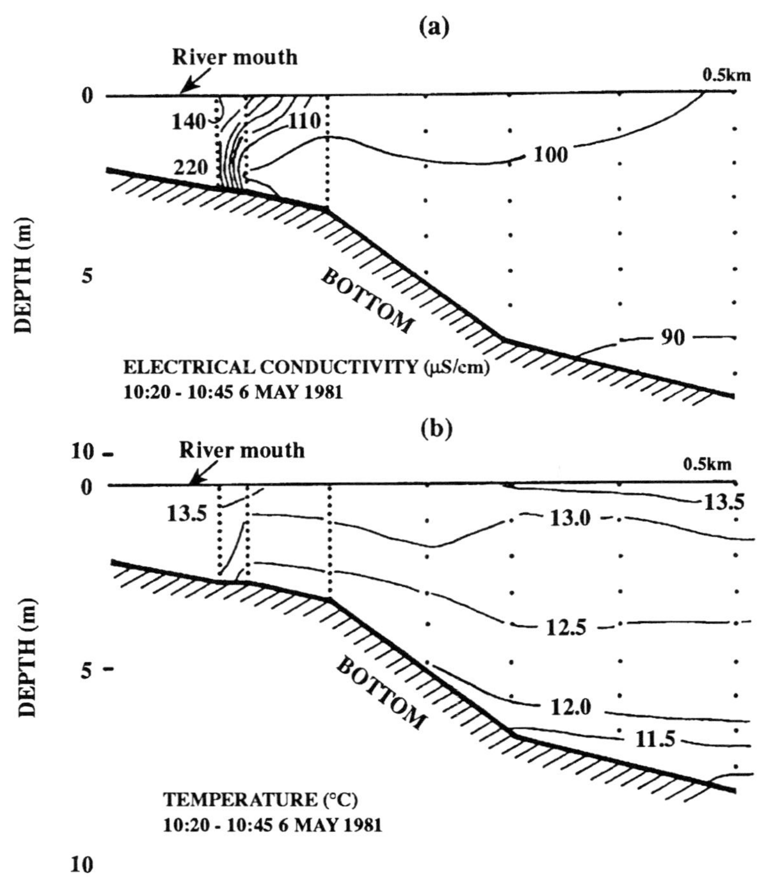


Figure 10.6. An example of the cross-sectional distribution of river water overlying the lake surface.

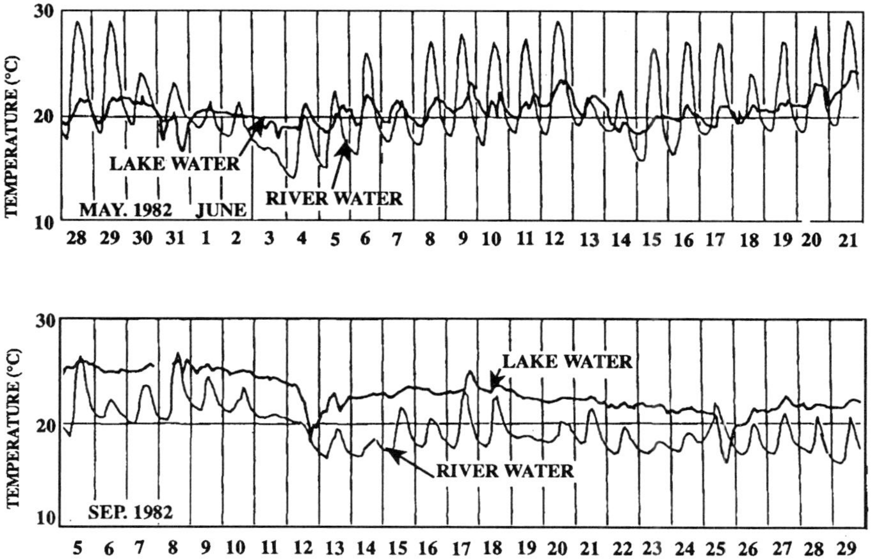


Figure 10.7. Comparison of temperature between the River Seri water and the surface layer water of Lake Biwa during a period of the warm (upper) and cold (lower) season.

# Distribution and Mixing of River Water

Figure 10.8 shows some examples of the horizontal distribution of the conductivity in the surface layer. A boat patrolled near the river mouth with sensors 25 cm below the lake surface, and the distribution of conductivity and temperature contours were estimated twice a day. Figure 10.8a shows the river water, forced by the lake-breeze, accumulating near the shore. After three hours, the lake breeze dropped causing a change in the conductivity distribution; the river water moved offshore (Figure 10.8b). It was calm the next morning and the river water mass was shaped like a tongue, reaching outwards into the lake (Figure 10.8c). Two hours later, the distribution changed and the tongue-shaped river water mass had split into two lenses (Figure 10.8d). Thus, Figure 10.8 illustrates that conductivity distribution (water mass distribution) is often complicated and highly variable, and depends on local wind conditions.

The distribution of conductivity averaged over repeated observations was found to be much simpler. Table 10.1 shows that the average values decrease regularly in the offshore region and the standard deviations, which reach a maximum value at Line Z (river mouth) also decreased with distance from the mouth.

The river water flowing into the lake basin after a heavy rain stays on the lake bottom for a long time with little vertical mixing. Table 10.2 is an example of the difference between the conductivity distribution before and after heavy rainfall. Rainfall was heavy on July 14, 15 and 16 of 1988 with a total precipitation of 169 mm. Automatic conductivity recorders were placed in the two layers (upper: 290 cm above bottom, lower: 25 cm above bottom) at Station B-3 during the period from 7 to 17 July 1988. The data collected every hour and averaged separately are presented in Table 10.2. The data show that conductivity increased from 99 μS/cm to 154 μS/cm in the lower layer (bottom layer), while it decreased in the upper layer over the same period. Table 10.2 implies that the river water is concentrated in the bottom layer with little upward mixing. The Richardson number computed at this station, was always much larger after heavy rain, indicating that vertical mixing was strongly suppressed. The conductivity decrease in the upper layer was apparently due to lake water moving onshore near the surface.

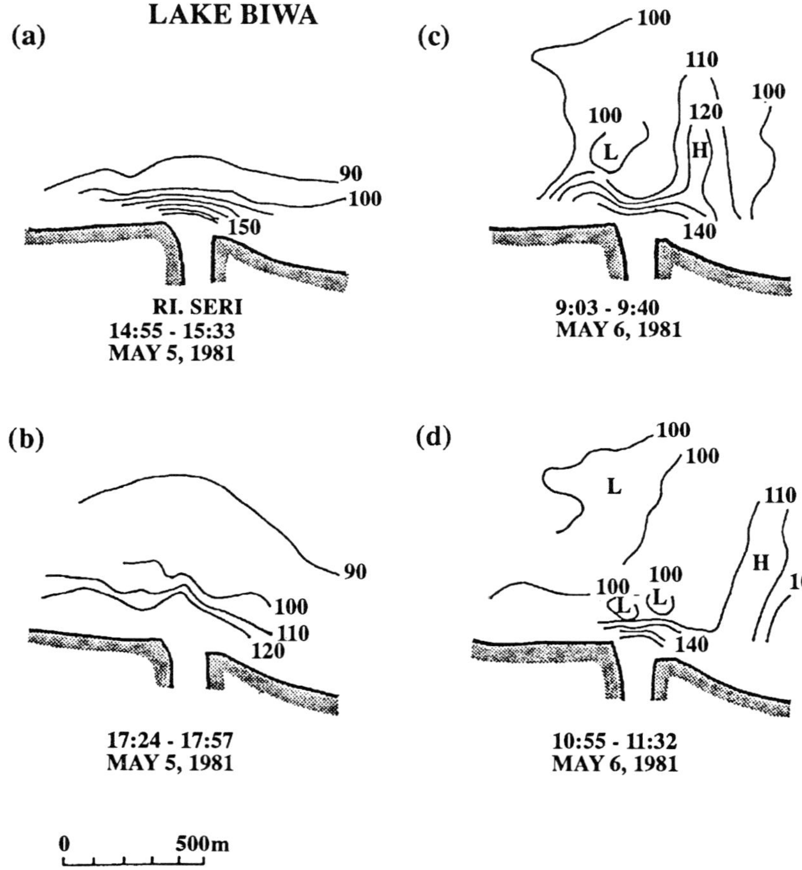


Figure 10.8. Examples of the changeable distribution of conductivity in the surface layer on successive observations.

TABLE 10.1. Horizontal distribution of conductivity and its standard deviation in the surface layer perpendicular to the shore line.

| Observation Line | Conductivity (μS/cm) Average | Deviation |
|---|---|---|
| Sta.Ikesu Br. | 205 | 6.5 |
| Line-Y | 196 | 16.2 |
| Line-Z | 155 | 24.4 |
| Line-A | 115 | 8.3 |
| Line-B | 108 | 4.2 |
| Line-C | 108 | 3.3 |
| Line-D | 106 | 3.5 |
| Line-E | 104 | 3.0 |

TABLE 10.2. Difference between distributions of averaged conductivities before and after heavy rain, July 1988

| Period | July 7 ~ 13 | July 14 ~ 17 |
|---|---|---|
| B3 Upper | 90 | 88 |
| B3 Lower | 99 | 154 |

(Upper: 290 cm above bottom Lower: 25 cm above bottom)

A water mass analysis was carried out using a T-C diagram (T: temperature, C: electrical conductivity). An example of the T-C diagram is shown in Figure 10.9 based on those data obtained immediately after the heavy rain in July 1988. The water masses can be classified into two types, one of which (shown in black circles) relates to Station I-3 situated 2.6 km from the river mouth, and the second (shown in open circles) to Stations A-3, B-3, C-3, D-3 and E-3 near the river mouth (locations shown in Figure 10.1). The former series consisted of pure lake water with a constant conductivity ($C = 95$), though with temperatures ranging from 13°C to 25°C as a result of thermal stratification. The latter series forms almost a straight line connecting point R ($C=140$, $T=16$) with point P ($C=95$, $T=23$-$25$). As points R and P represent original "river water type" and original "lake water type" in the surface layer, respectively, the water lying between R and P consists of mixed water masses of the two types. Nevertheless, this kind of mixed water type cannot be found within the triangle PQR, implying that river water did not mix with the cold ($T < 23°C$) deep lake water, but did mix with the warm ($T > 23°C$) surface layer water even when flowing along the bottom as a density current. Again this emphasizes the importance of analyzing mixing processes near the river mouth.

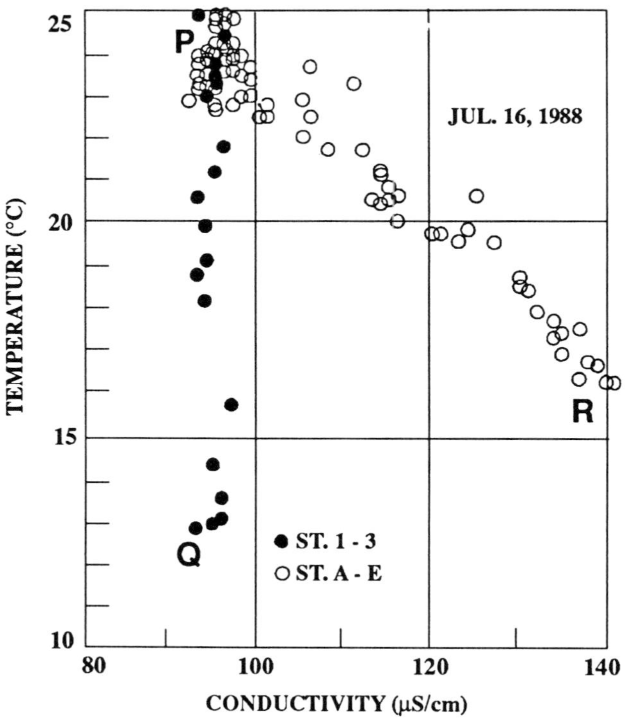


Figure 10.9. T-C diagram for water mass analysis in the offshore area immediately after heavy rain.

Longitudinal mixing along the river centerline can be assessed from the conductivity measurements made in the river. First, conductivity results were averaged vertically from surface to bottom and taken as a representative value at each individual station. Their longitudinal distributions along the river are shown separately in Figure 10.10a, which covers a distance of 400 m, from 200 m upstream to 200 m offshore. The individual distributions show the common trend of decreasing conductivity when moving offshore, but were somewhat irregular due to transient effects. Secondly, these data were again averaged over a period of four successive times, and the distribution of the averaged conductivity is shown in Figure 10.10b.

Using a box model (Figure 10.11) for an electrolyte budget we can write the following equation,

$$V\frac{\partial C}{\partial t} = QC_o - QC_1 - Ak\frac{\partial C}{\partial x} \tag{10.1}$$

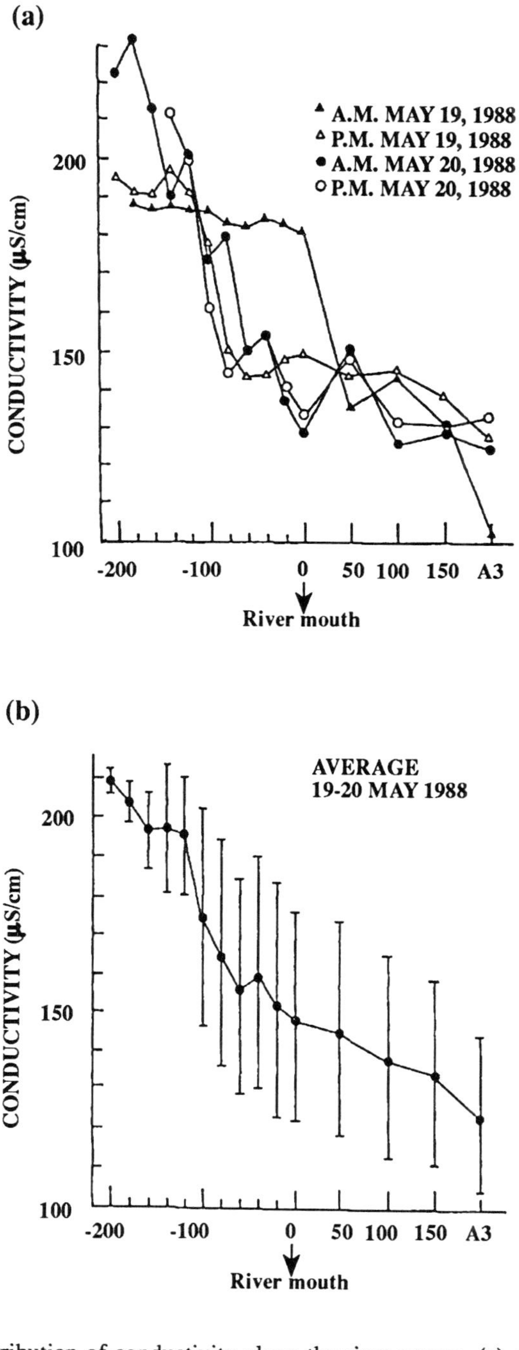


Figure 10.10. Distribution of conductivity along the river course. (a) averaged vertically from surface to bottom at each individual station, (b) averaged over four observations. Vertical bars represent standard deviations.

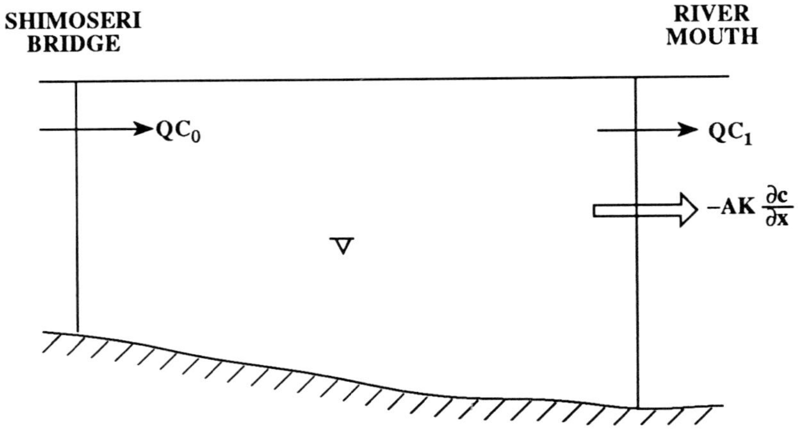


Figure 10.11. Box model for electrolyte budget near the river mouth.

with

| | | |
|---|---|---|
| C | : | concentration of electrolyte (proportional to conductivity) |
| $C_o$ | : | C at the left side boundary |
| $C_1$ | : | C at the right side boundary |
| Q | : | discharge |
| A | : | sectional area of river mouth |
| V | : | box volume |
| k | : | longitudinal coefficient of diffusion |

As we assumed that distribution in Figure 10.10b was steady, it follows that

$$\frac{\partial C}{\partial t} = 0 \tag{10.2}$$

Introducing the following figures, $\partial C/\partial x = 0$ at the left boundary, $\partial C/\partial x = -\ 0.10\ \mu S\ cm^{-1}\ m^{-1}$ the right boundary, $C_o = 205\ \mu S\ cm^{-1}$, $C_1 = 145\ \mu S\ cm^{-1}$, $Q = 80\ m^3\ s^{-1}$, $A = 140\ m^2$, we can obtain a value of k as follows:

$$k = 3.4 \times 10^4\ cm^2\ s^{-1} \tag{10.3}$$

(Fischer *et al.*, 1979). The relatively large coefficient of diffusion is the result of water exchange between river and lake, which was accompanied by a front.

Transport of electrolyte by advection and diffusion was 116 $\mu S\ cm^{-1}\ m^3\ s^{-1}$ and 48 $\mu S\ cm^{-1}\ m^3\ s^{-1}$ respectively. That implies a ratio of 2.4 : 1. The river water was thoroughly mixed with the lake water before leaving the river mouth. The vertical bars in Figure 10.10b denote the standard deviations of conductivity. The fact that they are large, from -100 m to +100 m, suggests the occasional appearance of a front.

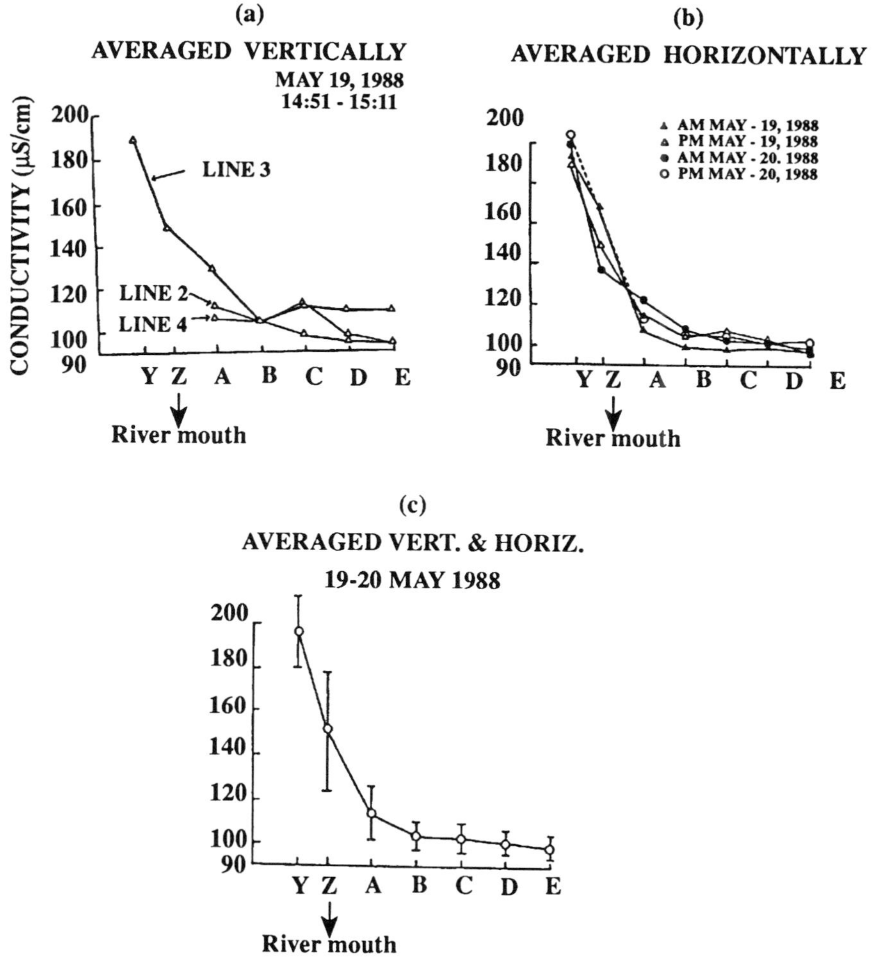


Figure 10.12. Distribution of conductivity in the direction perpendicular to the coastal line, (a) averaged vertically from surface to bottom at each individual station along three observation lines for the afternoon of May 19, 1988, (b) averaged over three observation lines at four times and (c) averaged over four successive times. Vertical bars represent standard deviations.

A method similar to that described above was applied to river water dispersion in the offshore region. Again the averaged conductivity shown in Figure 10.12 b was regular, in spite of the irregular distribution along an individual line (Figure 10.12a). A more regular distribution was found by averaging Figure 10.12b as shown in Figure 10.12c. Assuming a balance of electrolyte discharge from the river and also horizontally diffused across the vertical section parallel to the shore line, we estimated a horizontal diffusion coefficient at $2.8 \times 10^4\ cm^2\ s^{-1}$ for the case shown in Figure 10.12c.

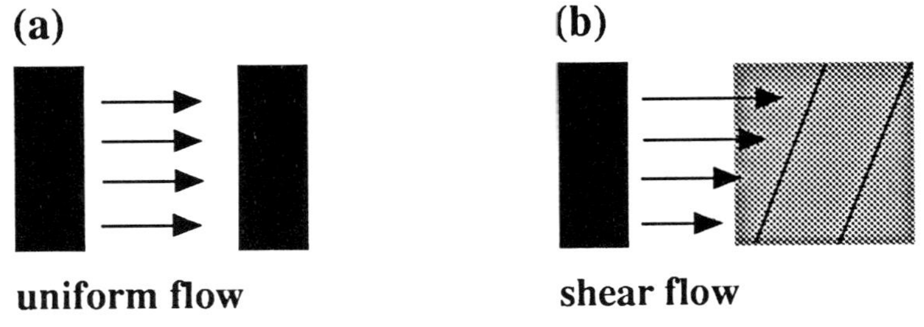


Figure 10.13. A conceptual diagram of the relationship between shear and diffusion.

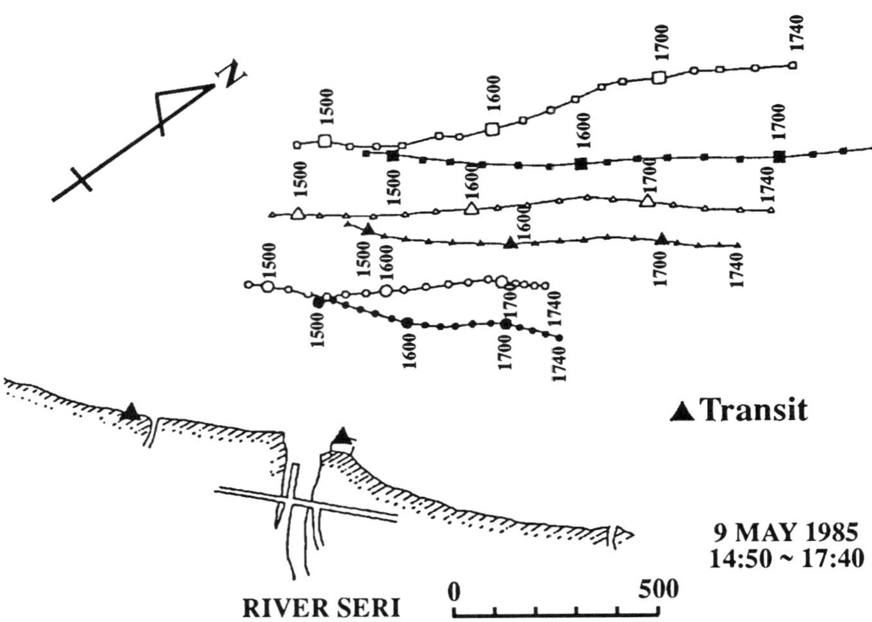


Figure 10.14. Drift trajectories at ten-minute intervals showing significant horizontal shear of velocity 500 m from the shore. Numbers denote observation times.

There are often large velocity shears of water current in the shore zone. The effect of shear upon diffusion is illustrated in Figure 10.13. On the assumption that no diffusion takes place in the current direction, a uniform current (Figure 10.13a) creates no diffusion, while a shear current makes a considerable contribution to diffusion as shown in Figure 10.13b. An example of a horizontal shear near the river mouth is shown in Figure 10.14, which shows drifter trajectories with the data recorded, at ten-minute intervals. There are remarkable differences in the current velocity between the offshore and near shore area. An example of vertical shear is shown in Figure 10.15, which depicts a time series of the velocity vectors in the upper and lower layers at an anchored station. There is almost always a considerable difference between the two layers. These strong shears explain the large horizontal diffusion coefficients.

## Summary

The distribution and dispersion of river water has been observed around the mouth of the River Seri flowing into Lake Biwa. The River Seri water had much higher electrical conductivity than Lake Biwa water and therefore conductivity could be used as a tracer for the inflowing river water.

Water fronts were observed at the boundary between river and lake. These underwent considerable deformation due to the influence of the lake/land breezes. The mixing of the two water masses was promoted by the front deformation causing a water mass exchange and consequently river water was already diluted before leaving the river mouth.

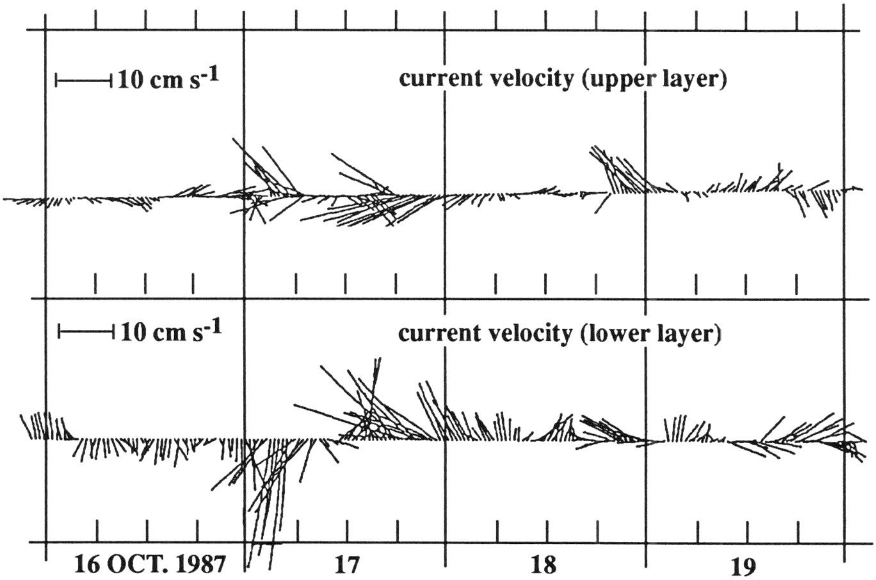


Figure 10.15. Time series of velocity vectors in the upper and lower layers showing vertical velocity shear. Data from an anchored station.

Whether the inflowing river overflows the lake surface or underflows along the bottom, depends on the temperature (density) difference between the water bodies under calm or weak wind conditions. In the case of an overflow the river water tended to mix with the lake water. Conversely, in the case of an underflow, particularly after heavy rain, river water tended to stay on the bottom with little vertical mixing. Water mass analysis using a temperature-conductivity diagram showed that immediately after heavy rains, river water mixed with rather shallow lake water around the river mouth and moved along the bottom with only little further mixing with the deep lake water.

Sudden rises of conductivity together with decreases in temperature were almost simultaneously recorded at some anchor stations located 1 km offshore. These changes are due to the appearance of a front advancing offshore and could be seen when drawing progressive vector diagrams. The actual distribution of conductivity was in general complicated near the river mouth; however, a simple or regular conductivity distribution could be extracted by averaging the individual values. Assuming that the actual distributions consisted of regular (steady) and irregular (fluctuating) distributions. The coefficient of longitudinal diffusion along the river was estimated $3.4 \times 10^4$ $cm^2$ $s^{-1}$.

## References

Anderson, E. R. and D. W. Pritchard, (1951) Physical limnology of Lake Mead. *U.S.Navy Electronics Laboratory*, San Diego,.153pp.

Carmack, E. C., C. B. J. Cray, C. H. Pharo, and R. J. Daley, (1979) Importance of lake-river interaction on seasonal patterns in the circulation of Kamloops Lake, British Colombia. *Limnol. Oceanogr.* 24: 634-644.

Churchill, J. H., (1985) Intrusions of outer shelf and slope water within the nearshore zone off Long Island, New York. *Limnol. Oceanogr.* 30: 972-986.

Fischer, H. B., (1980) Mixing processes on the Atlantic continental shelf, Cape Cod to Hatteras. *Limnol. Oceanogr.* 25: 114-125.

Fischer, H. B., E. J. List, R. C. Y. Koh, J. Imberger and N. B. Brooks, (1979) *Mixing in inland and coastal waters.* Academic Press, New York, pp 483.

Han, G. and T. Niedrauer, (1981) Hydrographic observations and mixing processes in the New York Bight, 1975-1977. *Limnol. Oceanogr.* 26: 1126-1141.

Ishitobi, Y. H. Kamiyama, K. Hayashi and M. Gomyodo, (1989) The tidal exchange in Lake Shinji under low discharge conditions. *Jpn. J. Limnol.* 50: 105-113.

Kao, T. W., (1977) Density currents and their applications. *J. Hydr. Div.*, Proc. ASCE, 103, HY5: 543-555.

Lindell, T., (1980) Hydrographic characteristics, in *Ecological Effect of Waste Water* edited by E. R. Welch,. pp 16-47.

Monismith, S.G., J. Imberger and M.L. Morison (1990) Convective motions in the sidearm of a small reservoir. *Limnol Oceanogr.* 35: 1676-1702.

# 11

# Sedimentation of Suspended Matter in the Mouth of the Ane River

Y. Okumura

## Introduction

Rivers and lakes have quite different response times to atmospheric forcing and this leads to complicated dynamics at the river mouth. Sediments originating in the catchment are transported through the mouth into a lake, the final sedimentation area. Coarser sediments are deposited around the river mouth, but finer particles are transported further into the lake and can be distributed over large areas. Very fine sediments do not settle, except under very calm conditions, and are capable of remaining in suspension for long periods of time.

Sediments can be classified by origin as either allochthonous (transported from the lake surroundings) or autochthonous (produced within the lake). Allochthonous sediments consist mainly of inorganic particles, while autochthonous sediments are essentially organic matter. Autochthonous sediments derive primarily from the productivity of the lake, while the amount of allochthonous sediment is largely determined by river discharges, as large amounts of sediment are injected into the lake during snow melts or heavy rains. Sediments that consist of very fine inorganic particles may cause turbidity and be transported to offshore regions. This chapter deals with the sedimentation of suspended sediments carried by the Ane River water into Lake Biwa.

Physical Processes in a Large Lake: Lake Biwa, Japan
Coastal and Estuarine Studies, Volume 48, Pages 119–127

# Methods of Investigation and Analysis

Field observations were conducted at the mouth of the Ane River, located in the northeastern section of Lake Biwa in Shiga Prefecture. The Ane River drains the second largest catchment region which supplies water to Lake Biwa. The geology of the drainage basin is mainly granitic, and it contains one of the areas of heaviest snowfall in Shiga Prefecture.

Sediments were sampled with sediment traps set on the lake bottom within 1.7 km of the river mouth. Water depths in the area ranged between 30 and 50 m (Figure 11.1). The sediment traps were constructed from acrylic resin cylinders with an aspect ratio of 3 (Gardner, 1980a,b; Håkanson and Jansson, 1983), with a height of 30 cm and diameter of 10 cm. Sediment traps were used in pairs with their openings set horizontally 2 m above the lake bottom. The observation period was about one month and the sediment traps did not have caps, as their retrieval was completed in calm conditions. After retrieval of the sediment traps, samples were collected in a plastic bottle, brought back to the laboratory, and dried under vacuum conditions. A daily accumulation rate (DAR) was calculated from each sample taken from the sediment traps and the DAR was given in units of amount of sample per unit area per day ($10^{-3}$ kg $m^2$ $day^{-1}$).

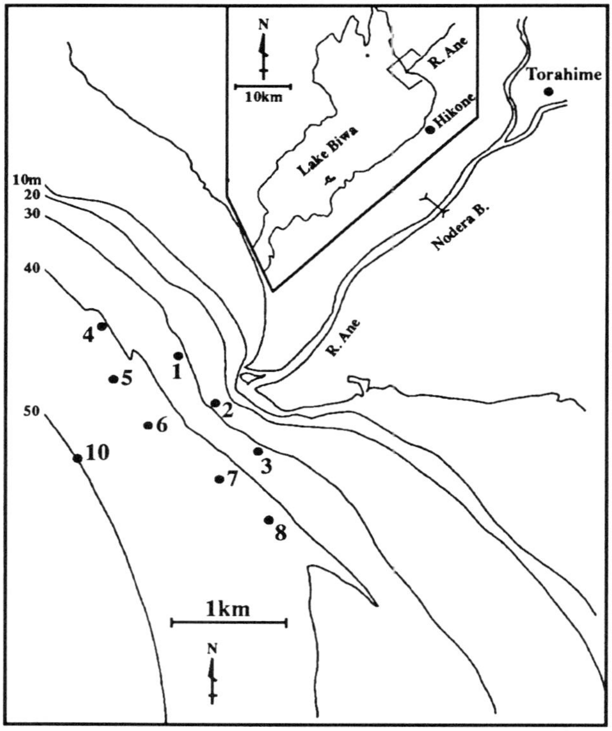


Figure 11.1. Observation stations in the mouth of the Ane River located in the North Basin of Lake Biwa.

# Geographic Distributions of DARs Around the River Mouth

Observations using sediment traps commenced in August 1984 and 22 deployments were completed by July 1988. The DARs and river discharges of the Ane River for each observation are shown in Table 11.1. River discharges were calculated from data recorded by a water level gauge located at the Nodera Bridge, 2.2 km upstream from the river mouth (Figure 11.1). Data obtained at Station 6 were the most complete and the geographic distributions of DARs obtained during several observation periods are shown in Figure 11.2. The area of each circle is proportional to the DAR. The upper row in Figure 11.2 shows the stagnation (thermal stratificiation period) period and the lower represents the circulation period (non-stratification period).

The geographic distributions of the DARs suggest that greater values tended to exist nearer the river mouth than in pelagic areas. The DAR values were also higher during the snow-melt flood season, and in the rainy season, although after heavy rain, DARs were large irrespective of season.

Naturally, wind direction and velocity must also be considered. During thermal stratification, wind impacts are limited to the epilimnion zone. After the fall overturn, the lake became isothermal, having a uniform temperature regardless of depth, with wind effects reaching the lake bottom. Sediment traps were set at a depth of about 40 m, where resuspension by wave action could be neglected. Nonetheless, it was necessary to consider the wind effect in cases of total offshore deposits in deep areas. Substances resuspended in coastal zones do not always deposit in the same place, and they may be transported offshore by wind-drift currents.

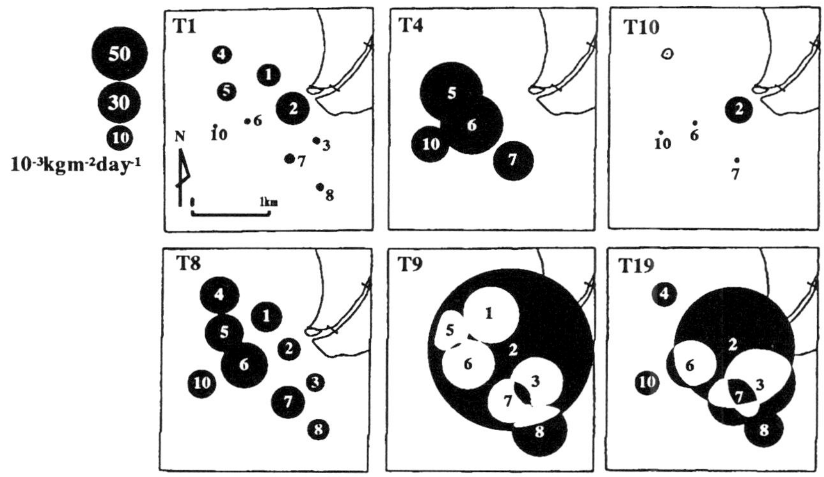


Figure 11.2. Examples of areal DAR (daily accumulation rate) distributions. The area of each circle is proportional to the DAR, upper plots are in the stagnation period; lower plots are in the circulation period.

TABLE 11.1 Daily accumulation rates and river discharges.

| Obs. No | Period | Duration (Days) | Station No. (water depth) 1 (32m) | 2 (26m) | 3 (33m) | 4 (41m) | 5 (43m) | 6 (43m) | 7 (42m) | 8 (42m) | 10 (50m) | Average River Discharge ($m^3 s^{-1}$) |
|---|---|---|---|---|---|---|---|---|---|---|---|---|
| T1 | 29 Aug - 19 Sep '84 | 22 | 8.88 | 18.9 | 1.72 | 5.52 | 4.24 | 1.38 | 2.64 | 2.14 | 0.09 | 4.45 |
| T2 | 19 Sep - 12 Oct '84 | 24 | 17.0 | 16.4 | 5.78 | ** | 5.43 | 4.77 | 2.53 | 2.07 | 0.73 | 0.036 |
| T3 | 7 June - 2 July '85 | 26 | | | | ** | ** | 21.1 | ** | ** | 8.59 | 43.79 |
| T4 | 2 July- 20 July '85 | 19 | | | | | 63.2 | 69.4 | 27.5 | | 23.7 | 90.76 |
| T5 | 12 Aug- 25 Aug '85 | 14 | 3.78 | 4.38 | 1.68 | | | 1.26 | | | | 0.445 |
| T6 | 24 April- 22 May '86 | 29 | | | | | | 3.12 | 0.19 | | | 16.25 |
| | 24 April- 30 May '86 | 37 | 1.36 | 0.28 | | | 1.73 | | | | | |
| T7 | 22 May- 12 June '86 | 22 | | | 4.54 | | | 1.54 | 2.48 | | | 1.87 |
| | 30 May- 12 Jun '86 | 14 | | 3.31 | | 1.86 | 2.70 | | | | | |
| T8 | 18 Dec- 22 Jan '87 | 36 | 15.3 | 8.56 | 5.00 | 25.6 | 23.5 | 34.9 | 19.5 | 7.31 | 8.84 | 7.65 |
| T9 | 23 Feb - 24 Mar '87 | 30 | 47.5 | 43.5 | 45.6 | | 24.7 | 43.7 | 34.6 | 54.4 | | 31.34 |
| T10 | 27 June - 23 July '87 | 27 | | 11.7 | | 2.91 | | 0.48 | 0.84 | | 2.24 | 11.24 |
| T11 | 23 July - 18 Aug '87 | 27 | | 4.86 | | 0.80 | | 0.39 | 0.04 | | | 2.78 |
| T12 | 18 Aug - 15 Sep '87 | 29 | | | | ** | | 1.90 | | | | 0.875 |
| T13 | 15 Sep - 14 Oct '87 | 29 | | | | 4.44 | | | | | | 0 |
| T14 | 14 Oct - 13 Nov 87 | 30 | | | | 12.1 | | 2.64 | | | | 5.97 |
| T15 | 13 Nov - 10 Dec 87 | 27 | | | | ** | | 13.9 | | | | 1.76 |
| T16 | 10 Dec - 6 Jan 88 | 28 | | | | 38.9 | | 25.9 | | | | 8.36 |
| T17 | 6 Jan - 4 Feb 88 | 30 | | | | 3.73 | | 20.9 | | | | 16.47 |
| T18 | 4 Feb - 2 Mar 88 | 27 | | | | 1.51 | | 9.98 | | | | 12.41 |
| T19 | 2 Mar - 29 Mar '88 | 28 | ** | 260 | 90.9 | 10.4 | ** | 38.7 | 45.4 | 25.8 | 5.53 | 46.38 |
| T20 | 29 Mar - 27 April '88 | 30 | | | | 2.13 | | 5.36 | | | | 11.98 |
| T21 | 27 April - 25 May '88 | 29 | | | | 9.32 | | 1.38 | | | | 10.36 |
| T22 | 25 May - 21 Jun '88 | 28 | | | | | | 8.0 | | | | 18.96 |
| T23 | 21 Jun - 18 July '88 | 28 | | | | | | 6.96 | | | | 25.52 |

** lost

# Relationships Between DARs, River Discharges, and Wind Velocity

The sediments caught by the sediment traps were mainly substances transported into the lake by the river. It was also necessary, however, to consider the contribution due to local resuspension caused by wave action. Okuda *et al.* (1986) have shown that resuspension is influenced not only by wave amplitude, but also by wave length. The character of the wind-driven waves depend on wind velocity, fetch, and duration. To assess the effects of wind-waves, the daily mean wind velocity was used as an index of wave-induced resuspension, with wind velocity data obtained by AMeDAS (Automated Meteorological Data Acquisition System) in Hikone. Figure 11.3 shows the relationships between DARs, river discharge, and wind velocities at Station 6. The abscissa is the river discharge of the Ane measured at Nodera Bridge, and the ordinate is the wind velocity at Hikone measured by AMeDAS. As before, the area of each circle in Figure 11.3 is proportional to the DAR.

Figure 11.3 indicates that DAR was a function of both river discharge and wind velocity. A multiple correlation coefficient was calculated with river discharge measurements at the Nodera Bridge and wind velocity as independent variables, and DAR as the dependent variable. The resulting multiple regression equation is:

$$DAR = -20.8 + 0.679Q + 9.62W$$

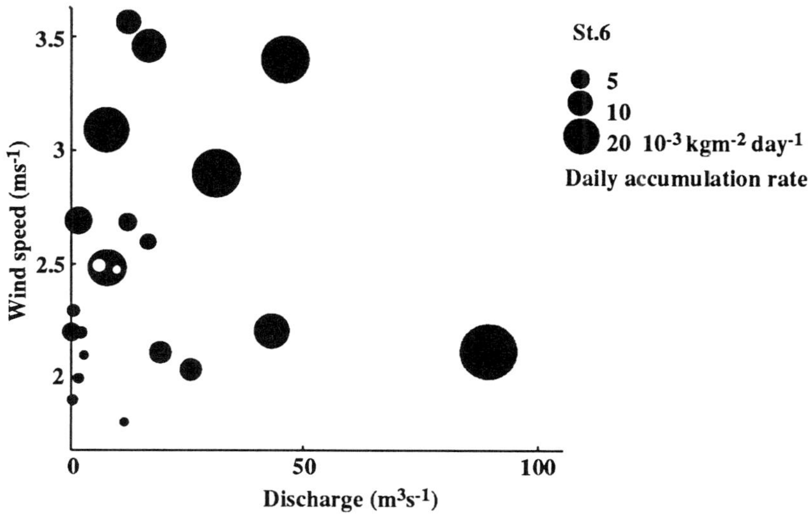


Figure 11.3. Relationships between DAR, river discharge, and wind speed. The area of each circle is proportional to the DAR, river discharges were measured at Nodera Bridge and wind speeds were observed in Hikone by AMeDAS.

where Q is the discharge of the Ane in $m^3 s^{-1}$, W is wind velocity in $ms^{-1}$ and DAR is in $10^{-3}$ kg $m^{-2}$ $day^{-1}$. The multiple correlation coefficient was R = 0.859. The correlations between DAR and river discharge is strong with the single correlation coefficient between DAR and river discharge being 0.813,; that between DAR and wind velocity 0.329. Thus, at Station 6, DAR is effected more by river discharge than by resuspension by wind.

The correlation coefficients at all observation stations are displayed in Figure 11.4. The thick letters are the station numbers and the three-digit numerals to the right of the Station numbers are the correlation coefficients. The upper numeral is the multiple correlation coefficient between the DAR, river discharge and wind velocity, the middle numeral is the single correlation coefficient between DAR and river discharge, and the lower numeral is the coefficient between DAR and wind velocity.

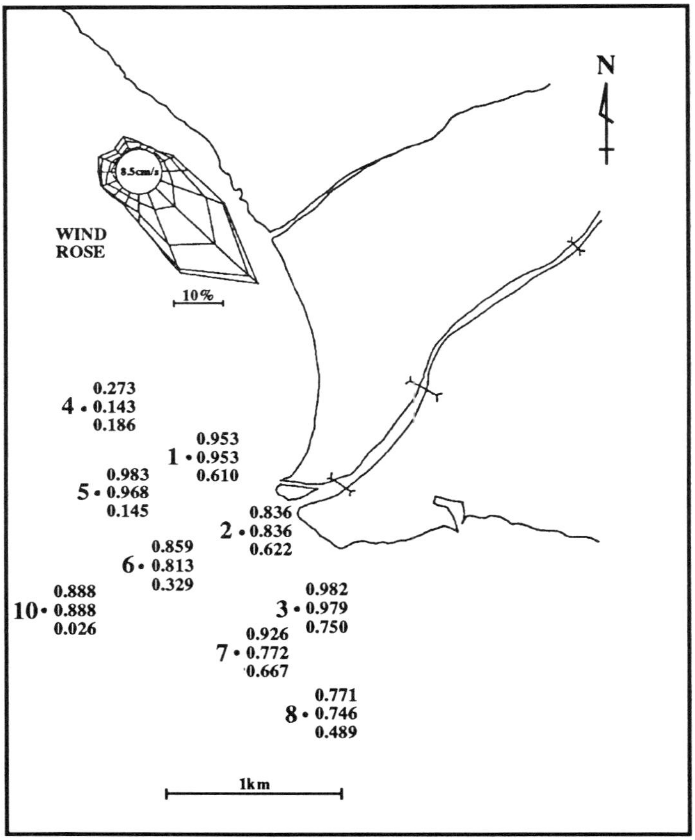


Figure 11.4. Areal distribution of correlation coefficients around the river mouth where upper coefficient is between DAR, discharge, and wind speed, middle between DAR and discharge; lower between DAR and wind speed.

The multiple correlation coefficients are extremely large except at Stations 4 and to a lesser extent at 8, these Stations being at the extremities of the survey region. As the multiple correlation coefficient at Station 4 is 0.273, neither river discharge nor the wind seem to influence sedimentation here. With the exception of Stations 4 and 8 the multiple correlation coefficients are very similar, but single correlation coefficients between DAR and wind velocity vary greatly from station to station. The single correlation coefficients between DAR and river discharge are both large and comparable in magnitude to the multiple correlation coefficients, implying the influence of the river extends uniformly over a wide range. The correlation between DAR and wind velocity was stronger near the river mouth, and became smaller offshore and was, for example, very weak at Station 10. In addition, the correlations were relatively stronger in the southern region of the river mouth, suggesting a greater degree of resuspension by wind closer to shore in the shallower water.

Annual weather statistics in Shiga Prefecture show that the daily mean wind velocity of seasonal northwest winds in winter is one and a half times greater than that of summer winds and, except during typhoons, the daily maximum wind velocity in winter is also larger than in summer. This suggests resuspension by wind waves is especially important in winter. Furthermore, currents around the river mouth tend to move south under the influence of the northwest seasonal wind in the winter months as observation records shown in the current rose in Figure 11.4. These effects should combine to yield stronger correlations between DAR and wind velocity extending toward the south of the river mouth in the winter period.

During the remainder of the year, the mean wind velocity is generally low, and the wind direction is quite variable. Though the currents around the river mouth are only weakly affected by wind, they have a tendency to flow northward due to the presence of a large-scale anti-clockwise gyre. In summary, sediment brought in by the river tends to be transported to the north in the summer and to the south in the winter. The influence of the wind is greatest near the shore and south from the river mouth and changes of DAR are much more influenced by river discharge than by wind velocity.

## DAR and Distance from the River Mouth

The relationship between DAR and distance from the river mouth was examined at points on a line extending along the centerline of the river. The points were at Stations 2, 6, and 10 and at an extra point further offshore where a sediment trap was set during one year from October 1983. The distances from the river mouth to these points were 0.35, 1.0, 1.65, and 2.0 km, respectively, and Figure 11.5 shows the relation between the distance and the DAR. A strong linear relation was observed when plotted logarithmically:

$$\mathrm{DAR} = 15.0 * L^{-1.6} \quad \text{(correlation coefficient } R = 0.994\text{)}$$

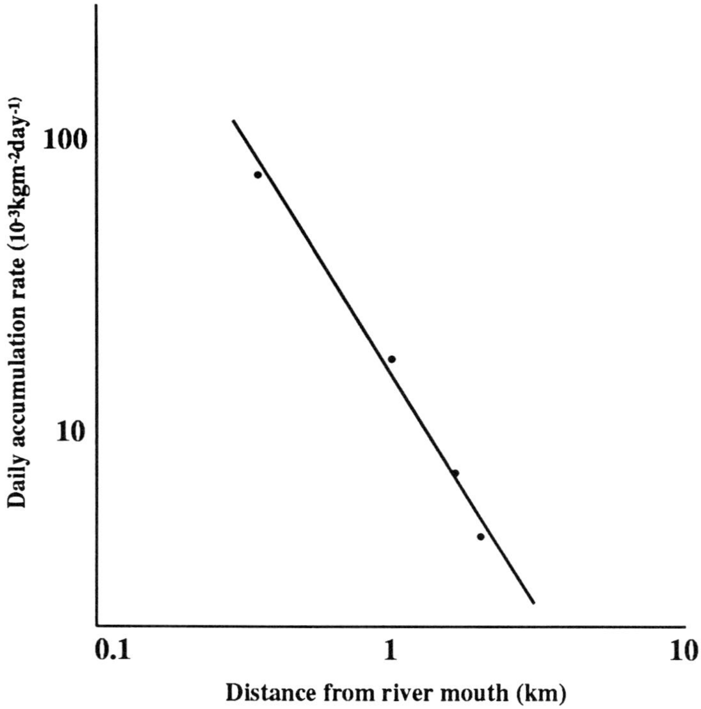


Figure 11.5. Relationships between the distance from the mouth of the Ane and mean DARs drawn on a logarithmic scale.

where L is the distance from the river mouth in km, and DAR is the average of the data from each station. The sediments transported into the lake by the river were thus deposited over a range of one to two kilometers from the mouth of the Ane.

## Conclusion

The sedimentation of suspended matter transported from the Ane River was investigated. Sediment traps were set within a distance of about 1.7 km from the river mouth. Multiple correlational analysis was carried out between DAR, river discharge, and wind velocity (taken as a measure of wind-wave resuspension effects). Strong correlations were demonstrated at all stations except Station 4. Single correlation coefficients were examined between DAR and river discharge and between DAR and wind velocity. The results indicated the effect of wind varied from point to point, the effect of the river discharge was much more uniform and dominant at almost all stations. Most deposition of river sediment occurred within one to two kilometers from the river mouth.

*Acknowledgments* . The author thanks A. Yamamoto, K. Namikoshi, and Y. Ohnishi of Osaka Electro-Communication University for their help with the investigation.

# References

Gardner, W. D., (1980a) Sediment trap dynamics and calibration: a laboratory evaluation. *J. Mar. Res.*, 38: 17-39.

Gardner, W. D., (1980b) Field assessment of sediment traps. *J. Mar. Res.*, 38: 41-52.

Håkanson, L. and M. Jansson, (1983) *Principles of Lake Sedimentology*. Springer-Verlag: 56-60 & 175.

Okuda, S., K. Yokoyama, K. Nishi, and M. Kumagai, (1986) Study on movement of lake bottom deposit, Part 1. Method and result of observation. *Annuals, Disas. Prev. Res. Inst., Kyoto Univ.*, 29 B-1: 341-353.

# 12

# Inflows Due to Snowmelt

M. Kumagai and H. Fushimi

## Introduction

The depletion of dissolved oxygen in the hypolimnion of the North Basin of Lake Biwa was first pointed out by Naka (1973). These results have been confirmed by the observations made by the Shiga Prefectural Institute of Public Health and Environmental Science. Eutrophication in Lake Biwa causes depletion of dissolved oxygen in the hypolimnion. When eutrophication becomes serious, the amount of organic matter in the lake increases and consumes much more oxygen when decomposing.

While certainly important, the effects of eutrophication may not fully explain the causes of oxygen depletion. Fushimi (1983) investigated the change in snow cover over a certain period of time and found a clear decrease, possibly related to the "greenhouse effect". Recently, Kumagai (1988) found that oxygen depletion was, in turn, also strongly affected by this decrease of snow cover in winter. Snowmelt in early spring carries significant levels of oxygen directly into the hypolimnion of Lake Biwa as a density current and thus determines the initial concentration of dissolved oxygen. Since a thermocline is present from May to October, the supply of oxygen from the epilimnion is suppressed by stratification, and oxygen is consumed in the hypolimnion and near the bottom by the decomposition of organic matter.

It is difficult to predict when the dissolved oxygen is completely exhausted, however, in assessing the temporal changes of dissolved oxygen in the hypolimnion, it is clear that the following factors must be taken into account:

Physical Processes in a Large Lake: Lake Biwa, Japan
Coastal and Estuarine Studies, Volume 48, Pages 129–139

(1) the amount of snow cover as an oxygen source,
(2) the amount of nutrient loading creating oxygen consumption,
(3) the flow dynamics in the hypolimnion as a transport mechanism for dissolved oxygen,
(4) the diffusion of oxygen through the thermocline as an oxygen flux,
(5) the consumption rate of dissolved oxygen in the hypolimnion and near the bottom, and
(6) the time and the strength of the overturn of the stratification and the effect on oxygen recovery.

If anoxic conditions occur in the hypolimnion and continue for a long time, nutrients are released from the bottom to the hypolimnion. There is little information from which to predict how long such a condition continues and whether or not the released nutrients are then transported from the hypolimnion to the epilimnion.

In this study, we focus on the dissolved oxygen supply by snowmelt and the effect on the concentration of dissolved oxygen in the hypolimnion of Lake Biwa. We then describe the results of observations of turbidity transport induced by snowmelt near the mouth of the Ane River as it flows into the Northern Basin of Lake Biwa.

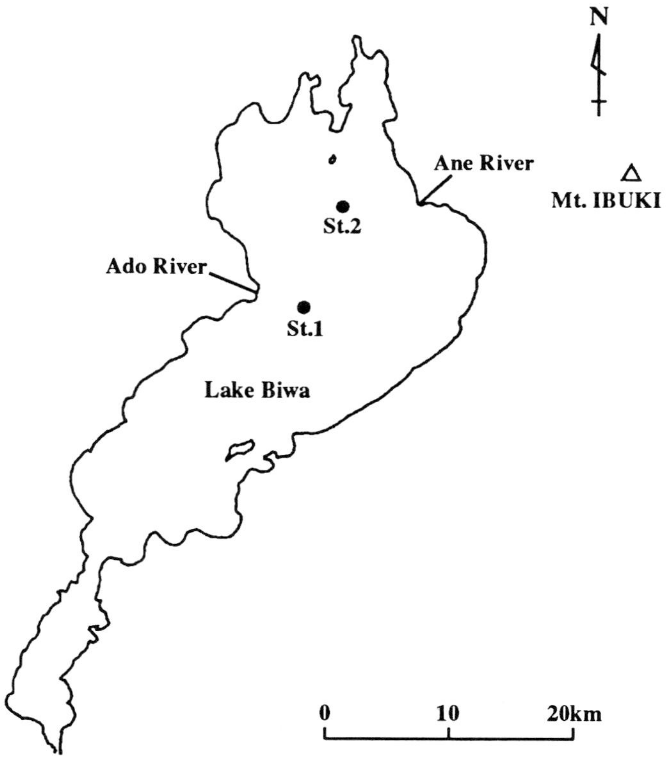


Figure 12.1. Map of observation stations in Lake Biwa.

## Data Collection

The water equivalents of snow cover were estimated using the degree of snow cover observed at 6 areas around Lake Biwa over the period 1968 to 1988. Dissolved oxygen data measured by the Shiga Prefectural Fisheries Experimental Station and the Shiga Prefectural Institute of Public Health and Environmental Science was also used. Turbidity, water temperature and discharge velocities from the Ane River in 1988 were measured using 6 sets of ACMs (Alec Current Meters) which contained a pressure gauge, x-y inclination meters, and a compass.

## Dissolved Oxygen Supply by Snowmelt

Figure 12.1 shows the observation stations, with Station 1 being the Shiga Prefectural Fisheries Experimental Station and Station 2, the Shiga Prefectural Institute of Public Health and Environmental Science Station. The approximate depths at the two stations were 80 m and 90 m, respectively. Observations were normally made twice a month.

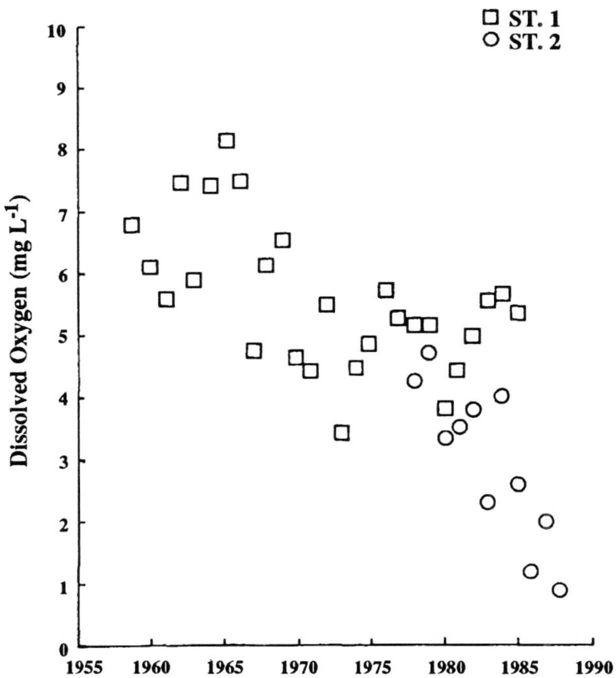


Figure 12.2. Temporal changes of the lowest annual dissolved oxygen concentrations in a hypolimnion.

Annual values of the minimum concentrations of dissolved oxygen in the hypolimnion at 1 m above the bottom are shown in Figure 12.2 for the period 1958 to 1988. It is clear there has been a steady decrease in the concentration of dissolved oxygen over this period and if the decrease continues at this rate, the possibility exists that anoxic conditions will occur in the hypolimnion by the end of this century.

Figure 12.3 shows the seasonal variation of dissolved oxygen at a depth of 80 m at Station 2 from 1984 to 1988. The seasonal cycle is similar for each year with dissolved oxygen at saturation concentration in April and decreasing gradually thereafter. In October or November, the concentration reaches a minimum, but overturning of the lake always causes a quick recovery. While dissolved oxygen is consumed during the decomposition of organic matter the value of the minimum in dissolved oxygen concentration will depend upon both the initial concentration at the onset of stratification and the timing of the autumn overturn.

Figure 12.4 shows the relationship between the water equivalent of the snow cover and the lowest annual level of dissolved oxygen at Station 1. It suggests that above a threshold value, in the range (0.5 to 1.0) x $10^9$ tons, dissolved oxygen levels are approximately constant at around 5 mg $L^{-1}$ while for lower levels of snow cover, the dissolved oxygen level decreases with decreasing snow cover. Similar effects can be seen in Figure 12.5 where dissolved oxygen measured 1 m above the bottom in April (the initial concentration) is almost at saturation levels, with most of the concentrations greater than 85%.

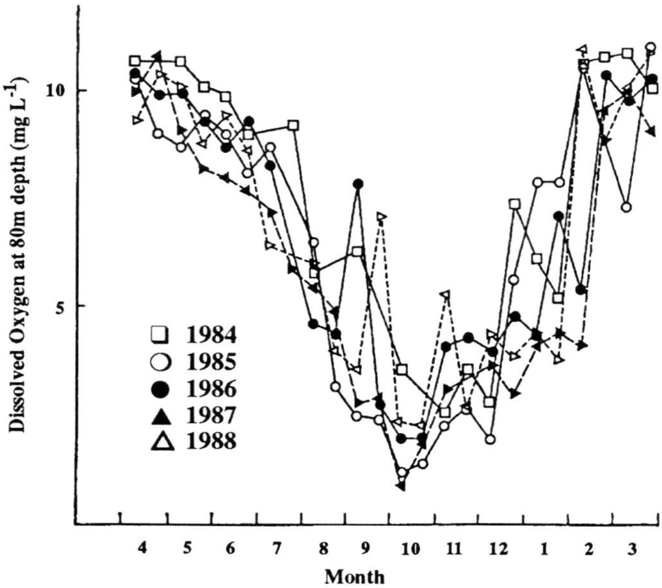


Figure 12.3. Temporal changes in dissolved oxygen concentrations at 80 m depth at Station 2.

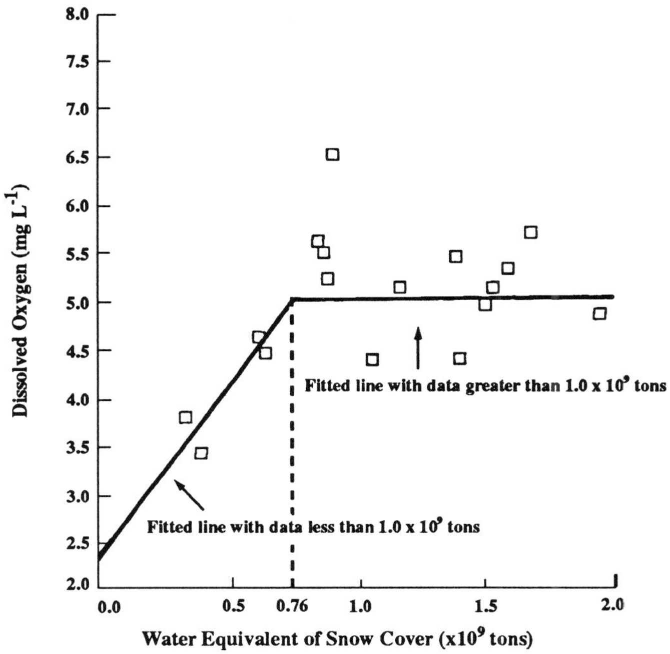


Figure 12.4. Relationship between water equivalent of snow cover and the lowest dissolved oxygen concentration in a hypolimnion at Station 1 in a year. Solid lines are determined with an optimization method.

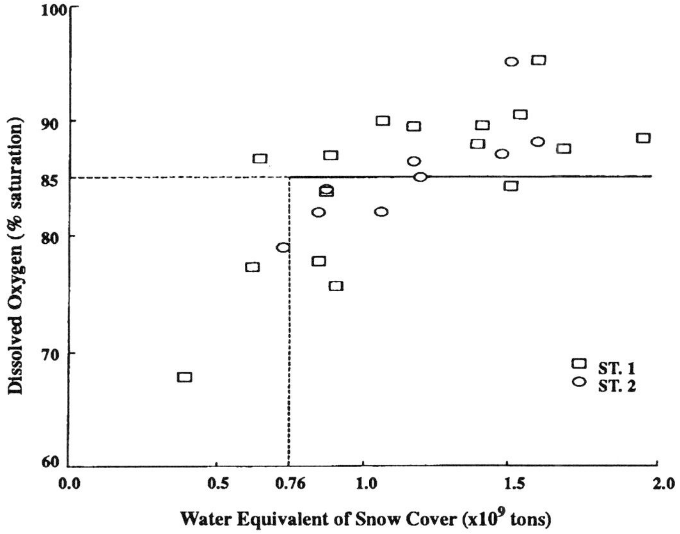


Figure 12.5. Relationship between water equivalent of snow cover and dissolved oxygen concentration at 1 m above the lake bottom in April.

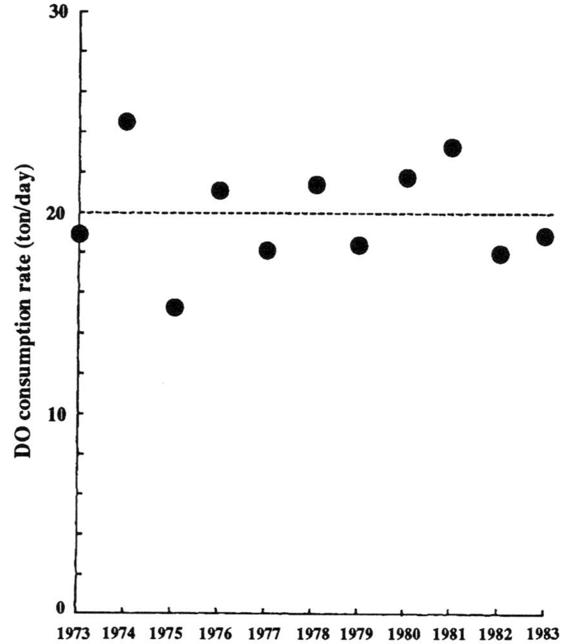


Figure 12.6. Temporal changes in the dissolved oxygen consumption rates in a layer below a depth of 80 m at Station 2.

In Figure 12.6, the consumption rate of dissolved oxygen in the layer below a depth of 80 m is shown. This rate is calculated as :

$$CR = ( DOA - DOL ) / DT ,$$

where CR is the consumption rate of dissolved oxygen, DOA the amount of dissolved oxygen in April, DOL, the amount of dissolved oxygen at the lowest concentration, and DT, the number of days between DOA and DOL. DOA and DOL were averaged over the region from 80 m to the lake bottom using field data from Station 1. The consumption rate (CR) was nearly constant at around 20 tons/day from 1973 to 1983, suggesting that the minimum dissolved oxygen concentration is not determined by consumption as much as by the initial condition and the length of stratification.

## Turbidity Supply by Snowmelt

In order to determine how snowmelt flows into Lake Biwa, field observations were carried out at the mouth of the Ane River in March 1988. The observation points, Stations A, B and C and their locations with respect to the river mouth, are shown in

Figure 12.7. A vertical cross-section of the observation Stations is shown in Figure 12.8, and the mooring methods of the ACMs are depicted in Figure 12.9. Only one ACM was moored at Station A and C while four ACMs were moored at Station B and the depth of each ACM is shown in Figure 12.9.

Turbidity was measured by the concentration of suspended solids with an ACM4M. Changes in turbidity over time at 50 cm above the bottom for Station A, Station B, and Station C are shown in Figure 12.10. The turbid density current caused by snowmelt was detected first at Station A on March 23, and subsequently at Station B and Station C. From these records, the speed of density current was estimated at 3.5 $cms^{-1}$ between Stations A and B and at about 1.3 $cms^{-1}$ between Stations B and C. The vertical structure of the turbidity current at four heights above the bottom at Station B is shown in Figure 12.11. A turbid density current passed through this station between the hours of 13:00 on March 22 and 7:00 on March 23. The thickness of the density current appears to be more than 430 cm. Taking advection into account and employing a spline interpolation, the length of the density current was predicted to be about 4,000 m and the thickness was about 20 m (Figure 12.12). The density current was defined as an area of high turbidity over 30 mg $L^{-1}$. If this density current reaches the deepest levels of Lake Biwa, it can be said that snowmelt is probably the most effective mechanism whereby dissolved oxygen is supplied to the hypolimnion.

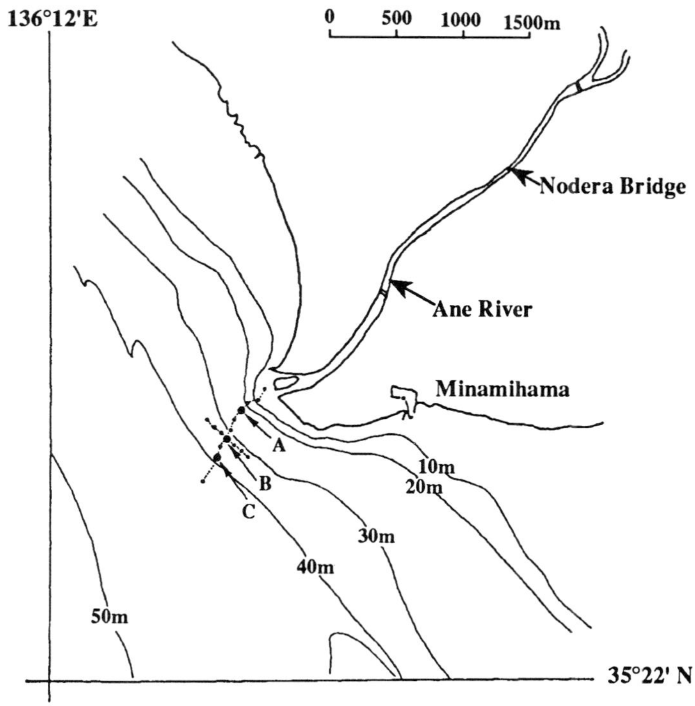


Figure 12.7. Observation stations near the Ane River.

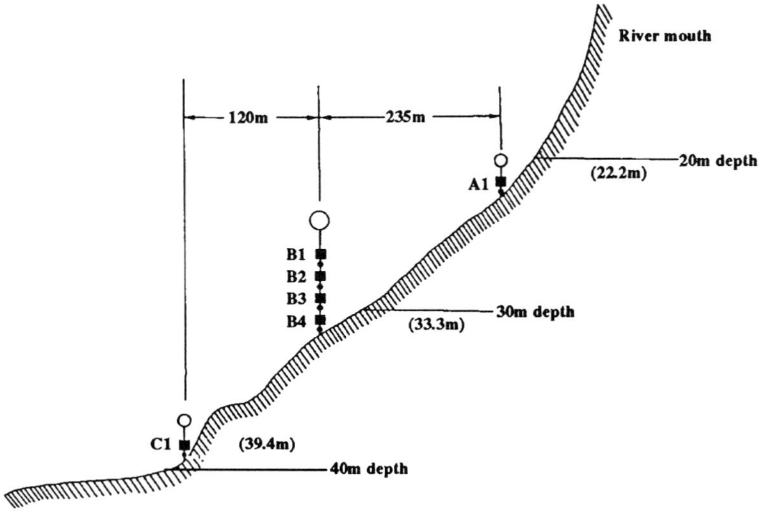


Figure 12.8. Positions of instruments set at the mouth of the Ane River.

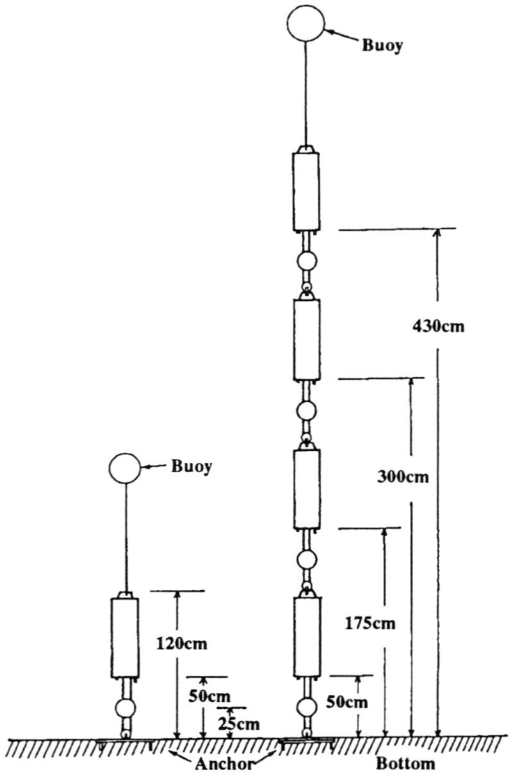


Figure 12.9. Mooring method for instruments.

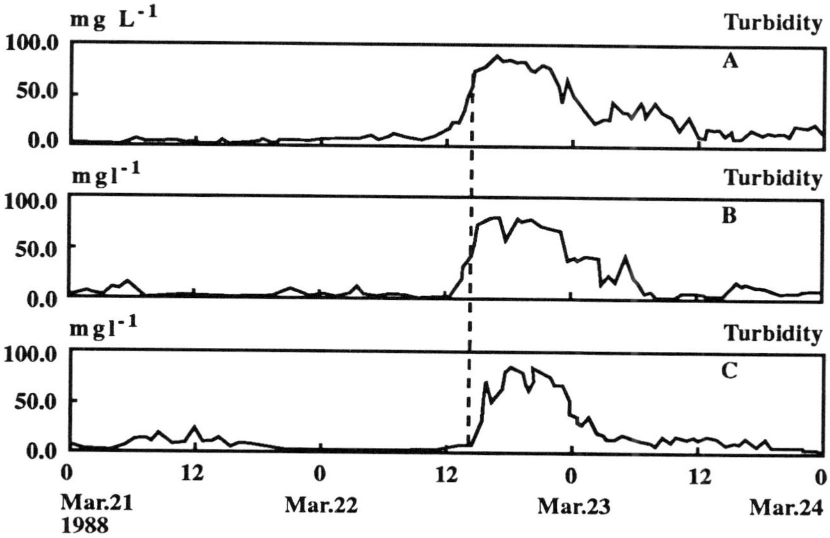


Figure 12.10. Temporal changes in turbidity near the lake bottom measured at Stations A, B, and C.

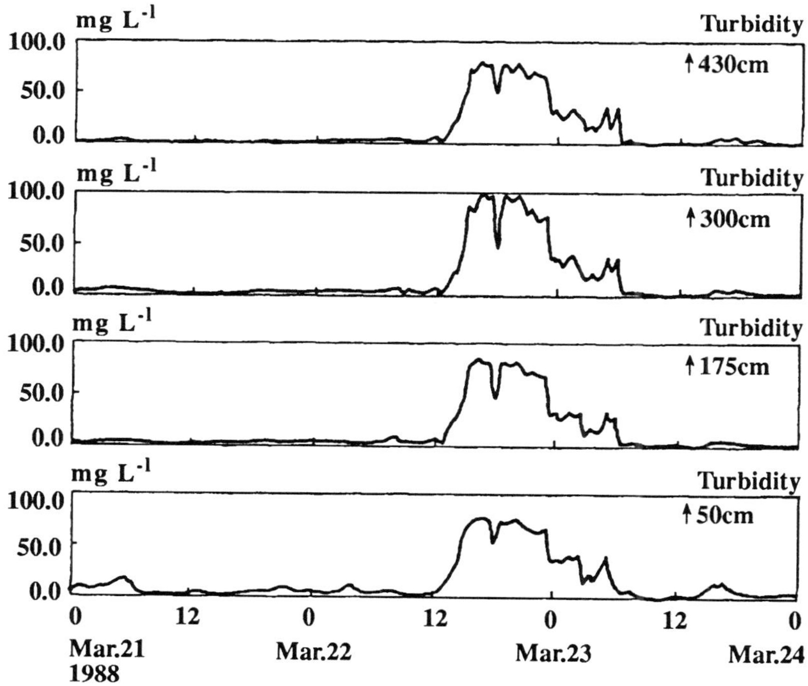


Figure 12.11. Temporal changes in turbidity measured at four levels at Station B.

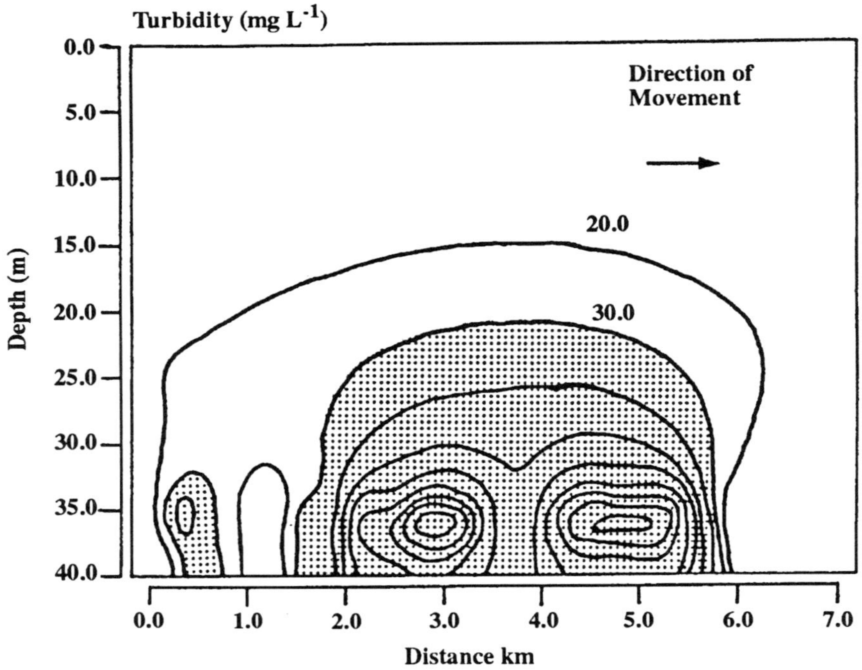


Figure 12.12. Spacial extent of turbid density current (shaded) induced by snowmelt.

# Conclusion

The decrease in the minimum annual dissolved oxygen concentrations in the hypolimnion of Lake Biwa is due not only to the effects of eutrophication but also to the decrease in snow cover. In particular, it seems that a critical value for the water equivalent of snow cover exists at about 0.5 x $10^9$ tons, below which the quantity of snow melt is proportional to the minimum observed oxygen levels at the bottom of the lake. In addition, the consumption rate of dissolved oxygen in Lake Biwa at a depth below 80 m is almost constant at a rate around 20 tons day$^{-1}$. Finally, a turbid density current of snowmelt in Lake Biwa was observed for the first time, with a length and thickness of about 4,000 m and 20 m, respectively.

*Acknowledgments.* The authors wish to express their sincere appreciation to Prof. T. Kira and Dr. Y. Oonishi at the Lake Biwa Research Institute for their helpful suggestions and for providing the opportunity to carry out field observations. We are also grateful to Prof. S. Okuda of Okayama Science University for helpful criticism, and to Ms. J. Okutani and Mr. T. Togo for their assistance in completing the manuscripts.

# References

Naka, K., (1973) Secular variation of oxygen change in the deep water of Lake Biwa. *Jap. J. Limnol.*, 34(1): 41-43.

Kumagai, M., (1988) Transport and sedimentation mechanism of suspended materials in Lake Biwa, in *The Front of Disaster Geomorphology*, edited by S. Okuda, pp 123-138.

Fushimi, H., (1983) Snow cover of the Lake Biwa catchment area - Structure of snow cover in the warm district - *Lake Biwa Res. Inst. Bulletin*, 2: 79-117.

# 13

# Recent Sedimentation Rates and Sedimentary Anomalies

H. Taishi

## Introduction

Lake Biwa is divided into a North Basin (Hokko) and a South Basin (Nanko). The North Basin is characterized by both north and south abysses located towards the western portion of the lake. The north abyss is enclosed by the 90 m depth contour and is deeper than the south abyss. The deepest point, at 104 m (marked KM in Figure 13.1), is in a local funnel-like region south of the north abyss.

The sedimentation rate is one of the most important and fundamental physical parameters in lake dynamics. The pattern of sedimentation rates in open water areas is strongly influenced by currents and wind/wave action (Håkanson and Jansson, 1983). Long periods of record are required to calculate sedimentation rates. Core samples of the sediment are useful in estimation of the rate. In general, sedimentation rates inferred from cores are related to paleo-hydrological processes in a lake, and sedimentation within the last 100 years is called recent. In this section, recent sedimentation rates are described using core samples from Lake Biwa. The related problems of sedimentary anomalies inferred from analyses of $^{210}Pb$ profiles of the core sediment are also discussed.

Physical Processes in a Large Lake: Lake Biwa, Japan
Coastal and Estuarine Studies, Volume 48, Pages 141–152

## Geographic Distribution of Recent Sedimentation

As summarized by Taishi *et al.* (1989) several authors have measured recent sedimentation rates from core samples from the bottom of Lake Biwa (Table 13.1 and Figure 13.1). Most of the cores fall in the recent sedimentation range and indicate sedimentation is in the order of 47-61 mg $cm^{-2}$ $yr^{-1}$.

TABLE 13.1 Recent sedimentation rates of core sediments estimated in Lake Biwa.

| Site*1 | Water Depth, m | Sed. Rate $mg \cdot cm^{-2} \cdot yr^{-1}$ | Applied Time Span or Depth | Method | Reference |
|---|---|---|---|---|---|
| P1 | 90 | 58 | 100 yr B.P | $^{210}Pb$ dating | Matsumoto (1975) |
| P2 | 70 | 48 | | | |
| P3 | 5 | 47 | | | |
| C1 | 10 | >41 | 0-5 cm | $^{137}Cs$ dating | Kamiyama *et al*., (1982) |
| C2 | 30 | 13 | | | |
| C3 | 70 | 37 | | | |
| C4 | 90 | 19 | | | |
| C5 | 73 | 39 | | | |
| A1 | 11 | 1.6*2 | 0-8 cm | based on vertical changes in grain density. | Yamamoto *et al*., (1982) |
| A5 | 43 | 1.3*2 | | | |
| A6 | 50 | 1.25*2 | | | |
| A7 | 56 | 1.1*2 | | | |
| A9 | 78 | 1.0*2 | | | |
| N1 | 70 | 33 | 100 yr B.P. | $^{210}Pb$ dating | Nakamura *et al*., (1986) |
| G1 | 31 | 103 | | $^{210}Pb$ dating | Taishi *et al*., (1989) |
| B3 | 52 | 56 | | | |
| B5 | 83 | 50 | | | |
| B6 | 88 | 47 | | | |
| B7 | 92 | 56 | | | |
| G2 | 93 | 61 | | | |
| KM | 104 | 55 | | | |
| G3 | 63 | 38 | | | |
| BW | 35 | 212 | | | |

*1 for the sites see Figure 13.1;

*2 relative values, setting the sedimentation rate in A9 at 1.0.

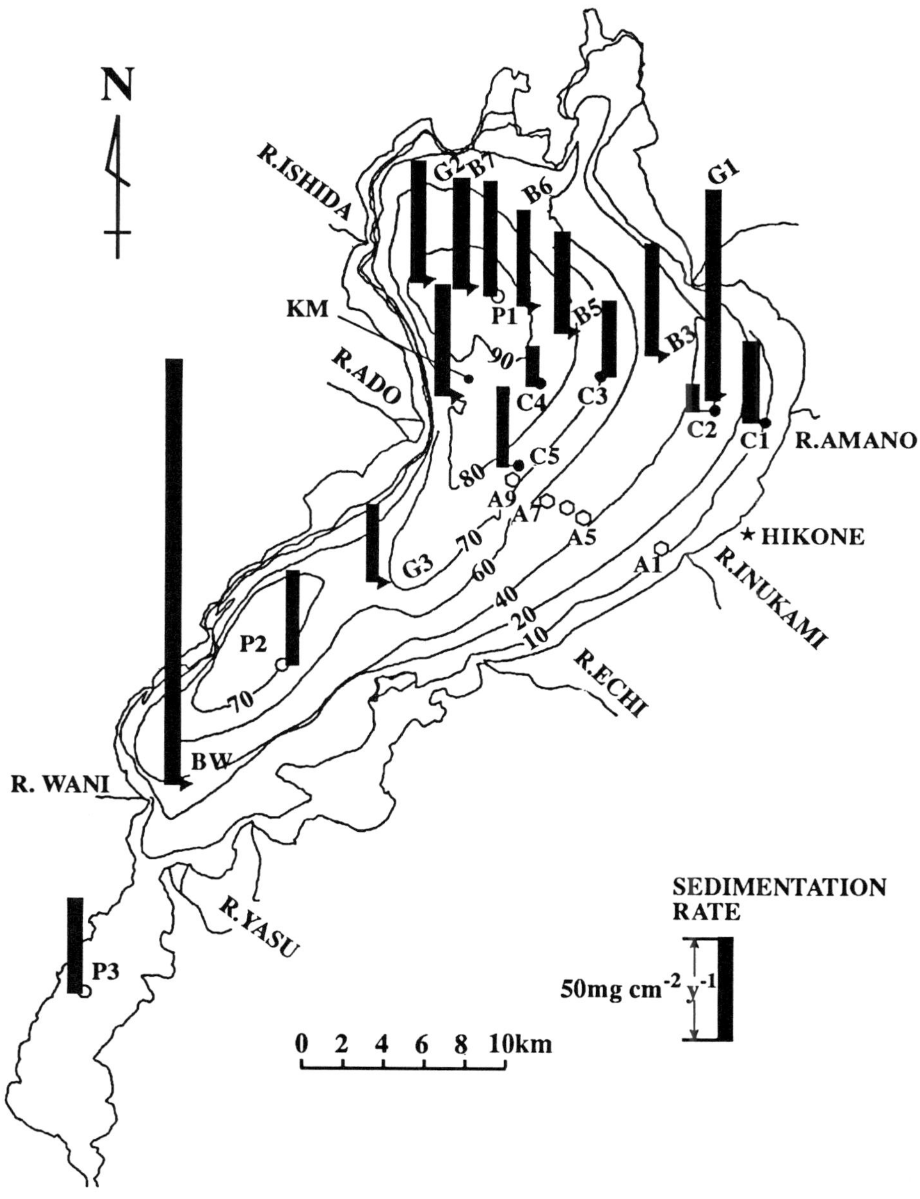


Figure 13.1. Sampling sites of core sediments and areal distribution of recent sedimentation in Lake Biwa. Rates estimated by Matsumoto, (1975), Kamiyama *et al.*, (1982), and Taishi *et al.*, (1989). Contours show water depth in metres.

Figure 13.1 shows the geographic distribution of recent sedimentation rates determined by Matsumoto (1975), Kamiyama *et al.* (1982), and Taishi *et al.* (1989). The data by Matsumoto (1975) and by Taishi *et al.* (1989) show that sedimentation rates in the main basin are larger in the shallow coastal zone (< ca. 40 m in depth) than in the deep central region (> ca. 40 m in depth). Based on an analysis of the granulometric properties of bottom core sediments along a transverse section Yamamoto *et al.* (1981, 1982) demonstrated that two discrete sedimentary environments exist in the main basin. One is a coastal zone influenced by one-way flows that tend, on average, to be away from the coast. The other is a central region affected by the gyre and occasional turbidity currents issuing from the river mouths. The boundary between them lies on the eastern side of the lake, between the 30 and 40 m isopleth.

Figure 13.2 shows the recent sedimentation rates against water depths using data from all authors in Table 13.1. Data by Yamamoto *et al.* (1982) are also plotted using a relative scale in sedimentation rate. The plot in Figure 13.2 shows that there are two definite relationships between the deposition rate and the depth in the central region. One is a roughly linear relationship provided by the eleven data points, i.e., P1, P2, C3, C5, N1, B5, B6, B7, G2, KM, and G3 (line I in Figure 13.2). These points lie in the central region with depths of more than 60 m (data point C4 is not included). On the other hand, the data points from A5, A6, A7 and A9 from Yamamoto *et al.* (1982) lie on a linear segment with a negative gradient (line II in Figure 13.2).

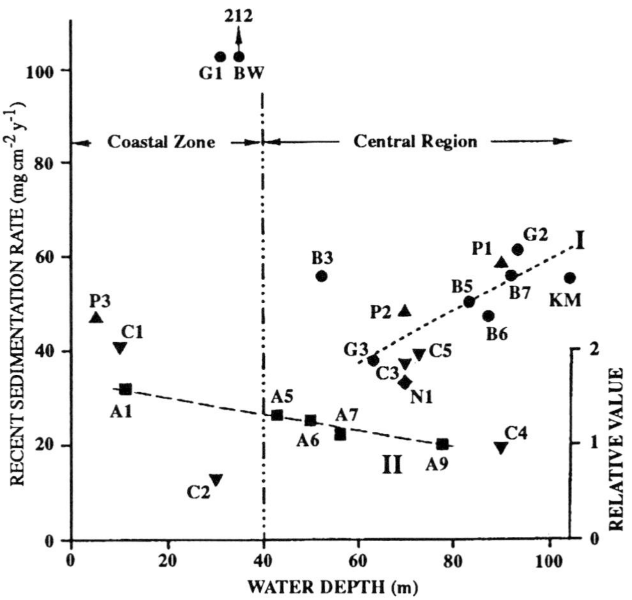


Figure 13.2. Plot of recent sedimentation rates to water depth, using data of Table 13.1. Line I shows a linear positive relationship provided by the eleven data points, i.e., P1, P2, C3, C5, N1, B5, B6, B7, G2, KM and G3 located in the western half of the central region at depth of more than 60 m. Line II shows a linear segment with a negative gradient based on the data points from A5, A6, A7 and A9 from Yamamoto *et al.*, (1982).

## Numerical Analysis of $^{210}$Pb Profiles

Most of the cores analyzed by Taishi *et al*. (1989) have two sections of uniform $^{210}$Pb activity interposed between normal layers in which the activity declined linearly. Figure 13.3 shows a typical $^{210}$Pb profile from core B5. It was suggested (Taishi *et al*., 1987a, 1989), that the uniform sections were associated with extraordinary sedimentation events, implying that recent sedimentation rates were drastically altered by extraordinary episodic sedimentation events during the past 100 years (Taishi *et al*., 1987a, 1989).

These $^{210}$Pb profiles were assumed to be formed by a combination of two independent sedimentation modes: the normal and accelerated sedimentation modes (curve I in Figure 13.4). Accelerated sedimentation is assumed to be caused by extraordinary sedimentation events during which large amounts of sediment quickly deposited on the bottom. Moreover, the profiles are assumed to be modified by mixing effects near the sediment-water interface, so that the $^{210}$Pb profiles are obtained as curves II (Figure 13.4). According to this interpretation, the observed $^{210}$ Pb profile from core B5 is composed of the three segments (U1, U2 and U3) and the two interval (A1 and A2), corresponding to normal and accelerated sedimentation modes, respectively (Figure 13.3).

The above conceptual interpretation of $^{210}$Pb profiles was quantitatively examined using a physical model of the $^{210}$Pb activity (Taishi *et al*., 1991). The model was based on a one-dimensional vertical advection-diffusion equation of $^{210}$Pb activity, and was formulated as:

$$\frac{\partial A}{\partial t} = E\frac{\partial^2 A}{\partial x^2} - \left(S_o - \frac{dE}{dx} - E\frac{dY}{dx}\right)\frac{\partial A}{\partial x} - \left\{\lambda - \frac{d}{dx}\left(E\frac{dY}{dx}\right)\right\}A, \tag{13.1}$$

where A (dpm $g^{-1}$) is the specific radioactivity of $^{210}$ Pb (dpm = disintegration per minute), t (yr) is the time, x (g $cm^{-2}$) is the mass based sediment depth below the sediment-water interface, E ($g^2\,cm^{-4}\,yr^{-1}$) is the mass based mixing parameter ($=\rho^2 D$), $\rho$ (g $cm^{-3}$) is the bulk sediment density, D ($cm^2\,yr^{-1}$) is the mixing coefficient, $S_o$ (g $cm^{-2}\,yr^{-1}$) is the mass-based sedimentation rate ($S_o = \rho V$), V (cm $yr^{-1}$) is the sediment rate, Y is the consolidation parameter ($=\ln\rho$ ), and $\lambda$ ($yr^{-1}$) is the radioactive decay constant. The boundary conditions are given as by:

$$\begin{aligned} F &= -E_o\frac{\partial A}{\partial x} + \left(S_o - E_o\frac{dY}{dx}\right)A, && \text{at } x = o,\\ \frac{dA}{dx} &= -\frac{\lambda A}{S_o}, && \text{at } x = -\infty, \end{aligned} \tag{13.2}$$

where F (dpm $cm^{-2}\,yr^{-1}$) is the flux of $^{210}$ Pb into the sediments and $E_o$ is the mass-based mixing parameter at the sediment-water interface. In the normal sedimentation mode,

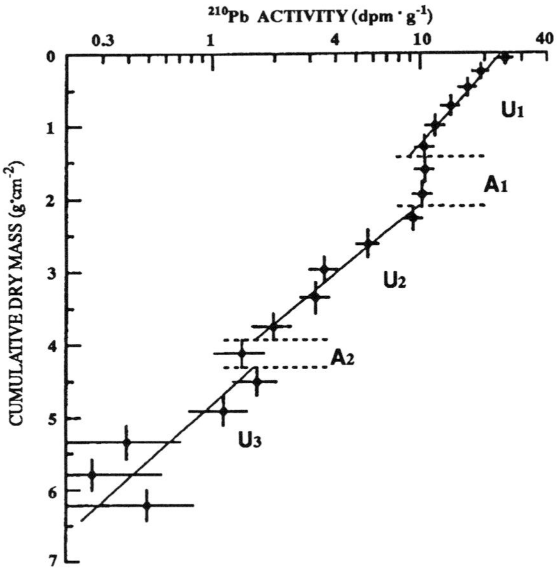


Figure 13.3. Activity profile of $^{210}Pb$ in core B5. U1, U2 and U3 are segments showing downward exponential reductions in activity. A1 and A2 are intervals of uniform activity associated with sedimentary anomalies. Modified by Taishi *et al.*, (1991).

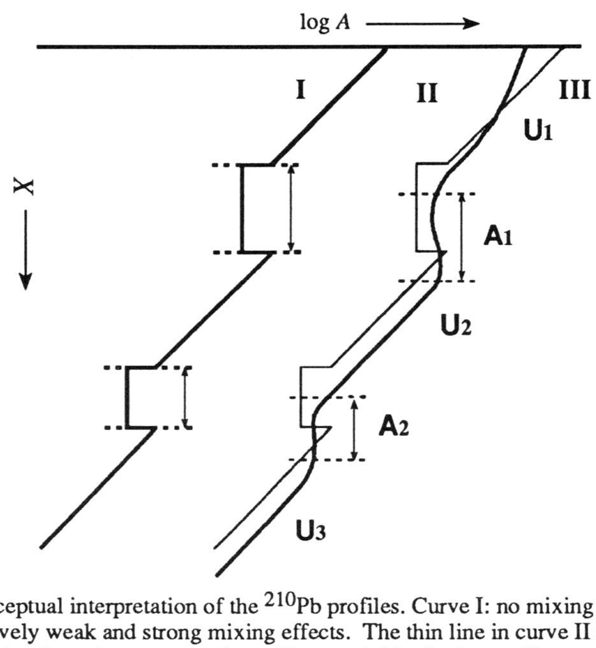


Figure 13.4. Conceptual interpretation of the $^{210}Pb$ profiles. Curve I: no mixing effects. Curve II: modified by relatively weak and strong mixing effects. The thin line in curve II is identical to that in curve I. Intervals where there seem to be uniform activities in the profile are marked with arrows.

the mixing effect in the top sediment layer and the consolidation effect of sediment layer are simultaneously considered in equation (1). The mixing coefficient, D, was assumed to have a Gaussian distribution over the depth, x: $D = D_0 \exp\{-x^2/(2X_m^2)\}$, where $D_0$ and $X_M$ are the mixing coefficients at the sediment-water interface and the effective mixing depth, respectively. The sedimentation rate $S_o$ was assumed to be constant. The solution reaches steady state and allows the consolidation parameter, Y (x), to be obtained from a polynominal best fit to the data in each of the cores.

The results of the numerical simulations for cores B5 and G2 taken from the central region of Lake Biwa are given in Figure 13.5. The calculated curves closely follow the observed $^{210}Pb$ profiles with intervals of uniform activity, but the model provides a poor fit for the top accelerated deposition layer. In general, numerical predictions proved satisfactory for cores from the central region, but application to cores from the coastal zone proved unsatisfactory. This is probably due to the fact that the latter coring sites were inactive sedimentary environments that were affected by river inflows, local currents, or both, so that simple one-dimensional models are not adequate.

Table 13.2 gives the parameters estimated for the cores from the central region. The values for the mixing coefficient $D_0$ (0.8-1.0 $cm^2\ yr^{-1}$) are comparable with those reported in previous studies (e.g. Robbins *et al.*, 1977; Christensen, 1982; Carpenter *et al.*, 1985; Kim & Burnett, 1988). The effective mixing depth $x_M$ (1.75-2.25 cm) is in agreement with data on benthic organisms in Lake Biwa (Narita, 1987). This suggests that minimum mixing occurs because of bioturbation near the lake bottom. The normal sedimentation rate, $S_o$, fell to a very narrow range between 30 and 40 mg $cm^{-2}\ yr^{-1}$. The maximum of 40 mg $cm^{-2}\ yr^{-1}$ occurred for core G2 where the water depth was 93 m and the minimum of 30 mg $cm^{-2}\ yr^{-1}$ was recorded at G3 where the depth was 63m. The thickness of accretion during the accelerated deposition, $T_A$, was largest in core G2 (1.0-1.5 g $cm^{-2}$) and smallest in cores B5 and KM (0.8 g $cm^{-2}$). Thus the two parameters, $S_0$ and $T_A$ , are in general, largest in the deep north abyss.

To estimate the contribution of the period of accelerated accretion, the mean sedimentation rate, $S_M$, is introduced as the ratio of the depth at the midpoint of the lower accelerated accretion and the time interval between its occurrence and the time of coring. Defined in this way, $S_M$ represents the mean sedimentation rate averaged over the past 90 years. The estimates of $S_M$ shown in Table 13.2 are almost the same as those previously obtained using the simple $^{210}$ Pb method (Taishi *et al.* (1989), Table 13.1). The contribution from the accelerated accretion, $S_A$, is expressed as the difference between $S_M$ and $S_o$ from Table 13.2. The values of $S_A$ are larger for core from the north abyss (cores B7 and G2), the shallow eastern part of the central region (core B3) and especially large for core G2, which was from the deepest abyss nearest to the coast. The contribution of the accelerated accretion when expresses as the ratio of $S_o/S_M$ was about 27-35%.

TABLE 13.2 Results of a numerical model of the $^{210}Pb$ data.

| Parameter | G1 | B3 | B5 | B6 | B7 | G2 | KM | G3 | BW |
|---|---|---|---|---|---|---|---|---|---|
| $D_0$ [$cm^2 \cdot yr^{-1}$] | 0.8 | 1.0 | 0.9 | 1.0 | 1.0 | 1.0 | 0.9 | 0.8 | 1.0 |
| $X_M$ [cm] | 3.0 | 2.0 | 1.75 | - | 2.25 | 2.25 | 1.75 | - | 8.75 |
| $S_o$ [$mg \cdot cm^{-2} \cdot yr^{-1}$] | 45 | 35 | 35 | 32 | 35 | 40 | 35 | 30 | 150 |
| $T_A$ [$g \cdot cm^{-2}$] in 1959 | 2.7 | 0.8 | 0.8 | 0.8 | 1.0 | 1.2 | 0.8 | 0.6 | 1.5 |
| $T_A$ [$g \cdot cm^{-2}$] in 1896 | - | 1.5 | 0.8 | 1.0 | 1.0 | 1.5 | 0.8 | 0.8 | - |
| $S_M$ [$mg \cdot cm^{-2} \cdot yr^{-1}$] | 93 | 52 | 49 | 46 | 52 | 62 | 48 | 41 | 220 |
| Water depth [m] | 31 | 52 | 83 | 88 | 92 | 93 | 104 | 63 | 35 |

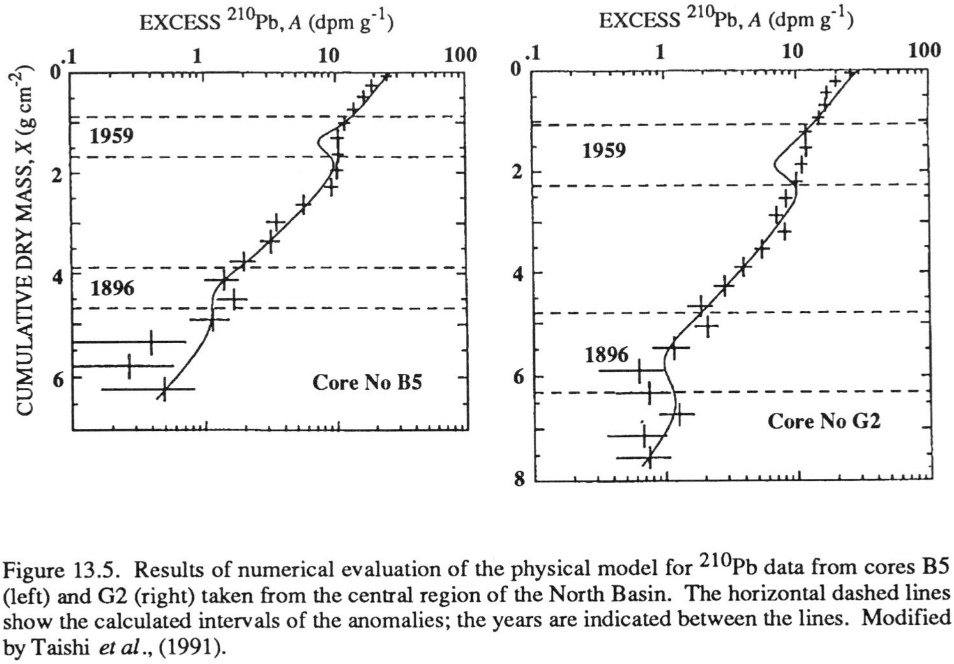


Figure 13.5. Results of numerical evaluation of the physical model for $^{210}Pb$ data from cores B5 (left) and G2 (right) taken from the central region of the North Basin. The horizontal dashed lines show the calculated intervals of the anomalies; the years are indicated between the lines. Modified by Taishi *et al*., (1991).

## Sedimentary Anomalies

The conceptual interpretation and numerical analysis of $^{210}Pb$ data suggests that two periods of accelerated sedimentation occurred, one around 1896 and another in 1959.

According to the meteorological data recorded over the past 90 years at Hikone City on the east coast of Lake Biwa (Figure 13.1), extraordinary rainfall occurred in 1896. A total precipitation of 1008 mm over 9 days (equivalent to 60% of the average annual precipitation) was recorded and the maximum daily precipitation was 597 mm. In 1959, the Ise-wan typhoon brought heavy rainfall (maximum daily precipitation, 271 mm) and strong winds (maximum velocity of 21.9 m $s^{-1}$) to the Lake Biwa area. These records therefore strongly suggest that the sedimentary anomalies are the result of episodic sedimentary transport to the lake mainly induced by exceptionally intense rainfall. Furthermore, waves generated by strong wind resuspend bottom sediments and produce a sedimentary flux from the shallow coastal areas of the lake to the interior (Okuda *et al.*, 1986; Kumagai, 1988).

Physical parameters of the sediments, including grain density, loss on ignition, water content, and grain size, were also measured for the samples from which the $^{210}Pb$ data was obtained (Taishi *et al.*, 1989). These secondary vertical variations of the physical parameters were used to detect possible signatures produced during the periods of rapid deposition (Taishi *et al.*, 1990). Inspection of these data show the following characteristics for the sediments deposited during the period of accelerated accretion: relatively high grain density, relatively low loss on ignition, relatively low water content, relatively coarse median size in the eastern part of the traverse (B3 and B5) and relatively fine size in the western part of the traverse (B7 and G2). Grain size analysis also showed that sediments coarser than sand made up less than 1% of the content, and there were no differences in content between the normal and anomalous layers. An analysis of the relationship between grain density and loss on ignition showed that the deviant variation in grain density was due to variation in the organic matter content (Taishi *et al.*, 1991). The sediments deposited during rapid accretion showed high grain density and low loss on ignition. These sediments were thus relatively mineral-rich compared with normal sediments; this is further evidence that these sediments originated during periods of intense river flow, which brought large amounts of sediment into Lake Biwa.

## Long-Term Sedimentation Rate and Sediment Focusing

In Lake Biwa, seismic sub-bottom profiling has revealed that prominent acoustic reflectors are correlated with volcanic ash layers of known age interposed with the sediment layers (Taishi *et al.*, 1987b, Inouchi 1987). As these ash layers range in age between 6 and 22 ky B.P., their sedimentation rates cover a time span of the order $10^3$-$10^4$ years. Figure 13.6 shows a change in long-term sedimentation rates based on

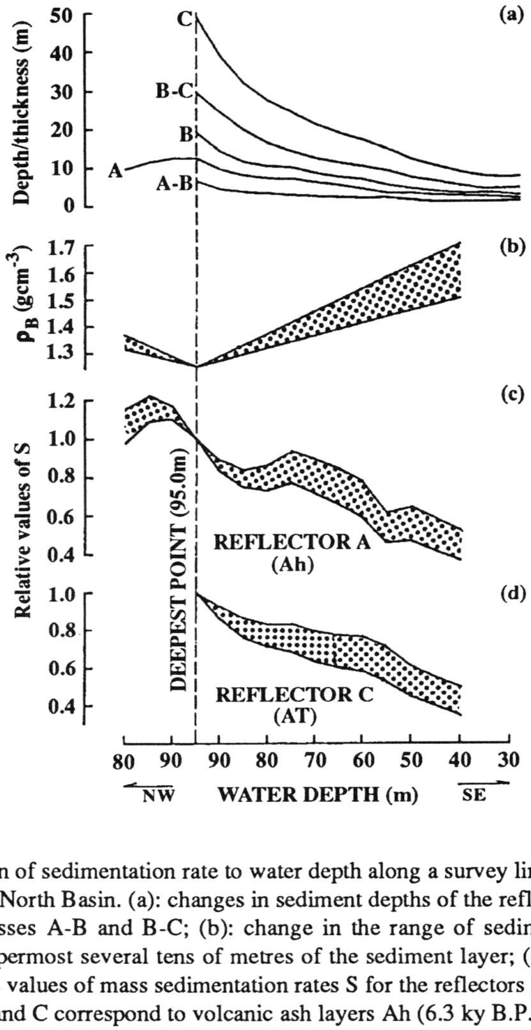


Figure 13.6. Relation of sedimentation rate to water depth along a survey line passing through the central region of the North Basin. (a): changes in sediment depths of the reflectors A, B and C and in sediment thicknesses A-B and B-C; (b): change in the range of sediment bulk density $\rho_B$ estimated for the uppermost several tens of metres of the sediment layer; (c) and (d): changes in the ranges of relative values of mass sedimentation rates S for the reflectors A and C, respectively. The reflectors A, B and C correspond to volcanic ash layers Ah (6.3 ky B.P.), U-Oki (9.3 ky B.P.) and AT (21-22 ky B.P.). Modified by Taishi *et al.*, (1987b)

seismic profiling along a survey line passing through the North Basin (Taishi *et al.*, 1987b).

The results show that sedimentation rates at depths greater than 40 m increase from the east coast to the deepest point in the central region. Such a phenomenon, whereby more sediments accumulate in the deeper parts, is called sediment focusing and is thought to be dominated by the configuration of the lake basin and the transportation-resuspension processes of sediments (Lehman, 1975; Hilton *et al.*, 1986). In some large lakes, Håkanson and Jansson (1983) demonstrated that there was a positive sedimentation rate versus water depth relationship in open water areas. The rate pattern is strongly influenced by currents and wind/wave action. Consequently, determining whether or not such sediment focusing has occurred in Lake Biwa during the recent time is very important in investigations of the accumulation and distribution processes of bottom sediments.

The recent sedimentation rate versus water depths relationships shown in Figure 13.2 indicate a focusing tendency towards the western part of the central region. However, the reverse tendency was observed in the eastern sector. These opposite tendencies may be viewed as anomalies in a general tendency for recent sedimentation rates to be small in the central part and large in the surrounding zones nearer to the coast. This would indicate that recent sedimentation rates are independent of water depths, and thus no sediment focusing has been effective in the central region of the lake during the past 100 years.

The contributions of the accelerated accretion during flooding obtained by analysis of $^{210}Pb$ profiles, revealed a tendency to be largest in the deepest north abyss and shallower in the eastern part of the central region. This suggests that these sediments are transported by inflow of river loads, and consequently more sediments have accumulated in the parts nearer the coast, where sediments are supplied.

# References


Carpenter, R., M. L. Peterson, and J. T. Bennett, (1985) $^{210}Pb$-derived sediment accumulation and mixing rates for the greater Puget Sound region. *Mar. Geol.* 64: 291-312

Christensen, E. R., (1982) A model for radionuclides in sediments influenced by mixing and compaction. *J. Geophy. Res.* 87: 566-572.

Håkanson, H. and M. Jansson, (1983) *Principles of Lake Sedimentology*. Springer-Verlag. pp. 316.

Hilton, J., J. P. Lishman, and P. V. Allen, (1986) The dominant processes of sediment distribution and focusing in a small, eutrophic, monomictic lake. *Limnol. Oceanogr.* 31: 125-133.

Inouchi, Y., (1987) Acoustic estimation method of sedimentation rate: case study in Lake Biwa. *Earth Science* 41: 231-241 (in Japanese).

Kamiyama, K., S. Okuda, and M. Koyama, (1982) Vertical distribution of 137Cs and its accumulation rate in lake sediments. *Jpn. J. Limnol.* 43: 35-38.

Kim, K. H. and W. C. Burnett, (1988) Accumulation and biological mixing of Peru margin sediments. *Mar. Geol.* 80: 181-194.

Kumagai, M., (1988) Predictive model for resuspension and deposition of bottom sediment in a lake. *Jpn. J. Limnol.* 49: 185-200.

Lehman, J. T., (1975) Reconstructing the rate of accumulation of lake sediment: the effect of sediment focusing. *Quat.. Res.* 5: 541-550.

Matsumoto, E., (1975) Accumulation rate of Lake Biwa sediments estimated by the $^{210}Pb$ method. *J. Geol. Soc. Japan* 81: 301-306 (in Japanese).

Nakamura, T., N. Nakai, M. Kimura, S. Kojima, and H. Maeda., (1986) Geological studies on radionuclides distributed in the bottom sediments of Lake Biwa. *J. Sedimentol. Soc. Japan* 25: 1-14.

Narita, T., (1987) Distribution of Tubificidae in the northern basin of Lake Biwa, p.121. Abstracts of 52-th Conf. *Jpn. Soc. Limnol.* (in Japanese).

Okuda, S., K. Yokoyama, K. Nishi, and M. Kumagai, (1986) Study on movement of lake bottom deposit: part 1. method and result of observation. *Annuals, Disas. Prev. Res. Inst., Kyoto Univ.* 29B(1): 341-353 (in Japanese).

Robbins, J. A., J. R. Krezoski, and S. C. Mozley, (1977) Radioactivity in sediments of the Great Lakes: post-depositional redistribution by deposit-feeding organisms. *Earth Planet. Sci. Lett.* 36: 325-333.

Taishi, H., S. Okuda, Y. Isozumi, and K. Yokoyama, (1987a) Estimations of sedimentation rates and sedimentary anomalies affected by abrupt meteorological events in the uppermost bottom sediments of Lake Biwa. *Res. Report Lake Biwa Res. Inst.* 86-A04:91-97 (in Japanese).

Taishi, H., Y. Inouchi, and T. Meguro, (1987b) Continuous seismic reflection sub-bottom profiling with "Uniboom" in the north basin of Lake Biwa: the structures of sedimentary layers and the origin of prominent acoustic reflectors. *Geophys. Bull. Hokkaido Univ.* 49: 251-268 (in Japanese).

Taishi, H., S. Okuda, Y. Isozumi, A. Hoshika, T. Shiki, K., Kashiwaya, K. Yokoyama, and Y. Inouchi, (1989) Radionuclide distributions and physical properties in core sediments from Lake Biwa. *Report of the Water Resources Research Center, Kyoto Univ.* 9:25-43 (in Japanese).

Taishi, H., S. Okuda, T. Shiki, and K. Kashiwaya, (1991) A sedimentary anomaly and the related sedimentation process. *Zeitschrift für Geomorphologie*, Supplementary Volume Bd.83:241-249.

Yamamoto, A., M. Higashihara, T. Meguro, and H. Taishi, (1981) On properties of bottom sediments and sedimentation processes in Lake Biwa, pp.73-93. In *Paleolimnology of Lake Biwa and the Japanese Pleistocene* 9, edited by S. Horie.

Yamamoto, A., M. Higashihara, T. Meguro, and H. Taishi, (1982) On properties of bottom sediments and sedimentation processes in Lake Biwa (II), pp.59-71. In *Paleolimnology of Lake Biwa and the Japanese Pleistocene* 10 edited by S. Horie.

# 14

# Bottom Sediments and Paleo-Hydrological Processes

A. Yamamoto

## Introduction

Lake Biwa, the largest lake in Japan, has a surface area of 674 $km^2$ and its drainage basin extends to 4.7 times the area of the lake. It is considered to be one of the oldest lakes in the world. Fine sediments carried in by rivers from the surrounding area are deposited in the central part of the lake where they have accumulated at a relatively high rate compared to other lakes. The bottom sediments, although containing remains of pollen and plankton, mainly consist of mineral grains produced by various erosional processes which have taken place in the surrounding drainage basin. As a consequence, these sediment grains have directly or indirectly experienced various meteorological, hydrological and hydraulic conditions and the grain size is a reflection of some of these environmental conditions. The geographic distribution in the grain size of surface bottom sediments indicates the present state of the lake sedimentation, while the vertical profile of the grain size of the bottom sediments indicates the temporal change in the sedimentary environment from the past to the present. In this paper, we shall discuss mainly the change in the paleo-hydrological conditions in Lake Biwa using results of grain size analysis from long core sediments.

Physical Processes in a Large Lake: Lake Biwa, Japan
Coastal and Estuarine Studies, Volume 48, Pages 153–167

## Areal Distribution of Grain Size on the Lake Bottom

In Lake Biwa, the first attempt to get the areal distribution of the grain size on the bottom was conducted by Morikawa *et al.* (1957) who constructed sediment distribution maps. The distribution maps were later revised by the author with the addition of new data and presented in terms of Inman's grain size parameters. An example of the maps is shown in Figure 14.1 for the North Basin of the lake. Recently, Kamitani (1988) completed a new distribution map based on the grain size analysis of 164 bottom samples taken from points along a 2 km grid.

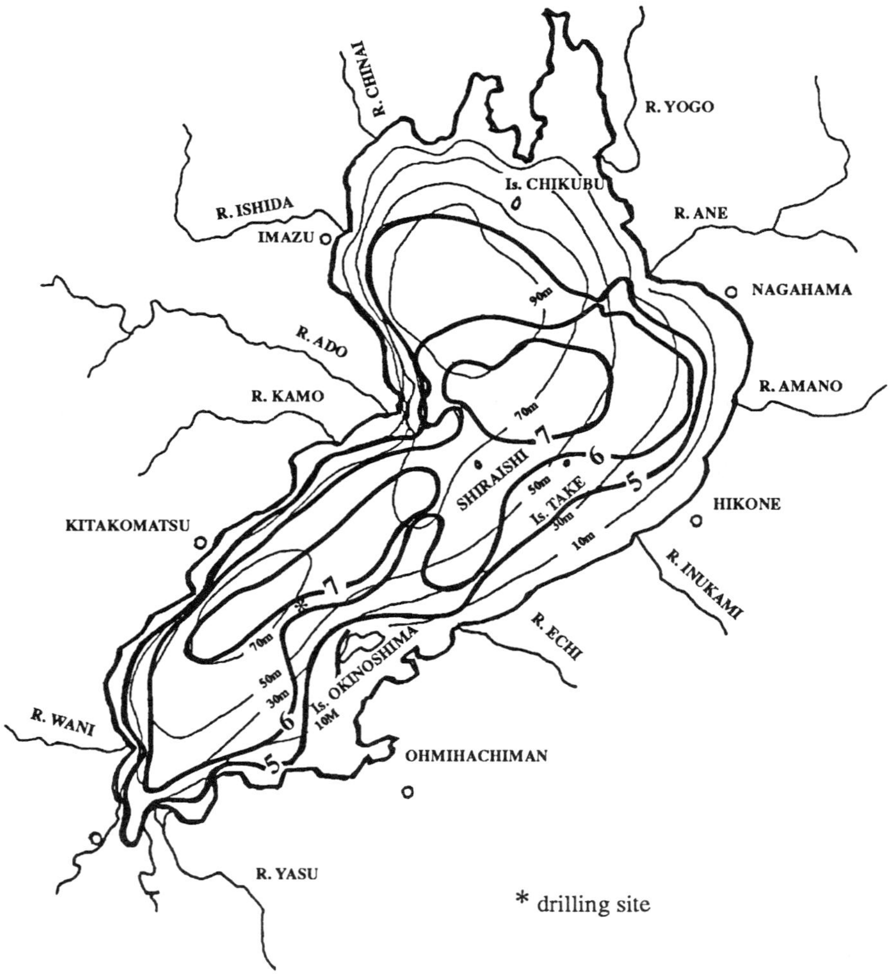


Figure 14.1. Areal distribution of $Md_\phi$ in Lake Biwa (revised after Morikawa *et al.*, 1957) and drilling site (*). Thin lines indicate bottom contours in metres.

From consideration of detailed grain size analysis of short cores (not exceeding 20 cm in length) from surface sediments, the lake bottom can be divided into two regions : the central depositional region with a water depth exceeding 50 m, and the shore region, shallower than 50 m, which includes the true shore area shallower than 10 m (Yamamoto *et al.*, 1982). Cores were taken at 15 sites along a transverse survey line connecting the mouths of Inukami River and the Ado River. In general, there is a good correlation between the phi-median diameter ($Md_\phi$) and the phi-skewness measure ($\alpha_\phi$) for surface sediments, $Md_\phi$ becoming finer as $\alpha_\phi$ becomes less skewed with increasing distance from the shore. The $Md_\phi$ - $\alpha_\phi$ diagram (Figure 14.2) reveals that sediments which have settled at the central depositional region deeper than 50 m occupy the definite domains "A and B", whereas the sediments along the shore region occupy domains "C and D". The domain C is associated with an intermediate zone between the shore and mid-lake, while the domain D belongs purely to the shore zone shallower than 10 m. The domain A is usually occupied by uniform fine sediments.

## Grain Size of Core Sediments and Paleo-Precipitational Change During the Past Half Million Years

At a site near the mid-point between Kitakomatsu and Okino-shima Island in the central part of Lake Biwa (Figure 14.1), two long cores of bottom sediments were taken in water of depth 68.4 m by S. Horie. The first long core was 200m, drilled in 1970. In 1982-83, a 1,400m core was drilled which was the longest (deepest) core to have been taken from any lake in the world. Both cores were investigated by an interdisciplinary team and yielded a large quantity of useful information (e.g., Horie, 1972-83). A 12 m core was also taken at virtually the same site in 1967.

The 12 m core covered the entire Holocene and some paleoenvironmental analyses of the core were carried out (Horie *et al.*, 1971). The grain size variations obtained from this core are shown in Figure 14.3. The sequence in the median shows that finer particles were deposited during the last glacial period (ca.25-12 kys B.P.), whereas coarser particles were deposited after this period. The figure also indicates that the grain particles became a little finer after the climatic optimum (ca.9-3 kys B.P.). The variation in sorting (i.e. phi-deviation measure $\sigma_\phi$) appears to have a period of ca.1,400 years and there are two peaks in the sequence of very fine grains during the period of deposition of relatively coarse grains since the last glacial period (Yamamoto, 1977).

The 200 m core, which consisted mainly of uniform fine muds, was analyzed by using 780 pieces of 5 cm long samples taken out at intervals of 25 cm. For every sample, grain density and grain size distributions were measured. To analyze the grain size, wet sieving (for grains coarser than 44 μm) and settling analysis using the photo-extinction method were employed. An age scale was constructed on the bases of the consolidation model, where a time sequence of bulk densities, three $^{14}C$ dates and six fission-track dates were used to estimate the scale. The age of the bottom of the 200 m core was about 0.5 my B.P. (Yamamoto *et.al.*, 1973). Vertical variations in the grain density and some of Inman's grain size parameters are shown in Figure 14.4 (Yamamoto *et* al., 1973).

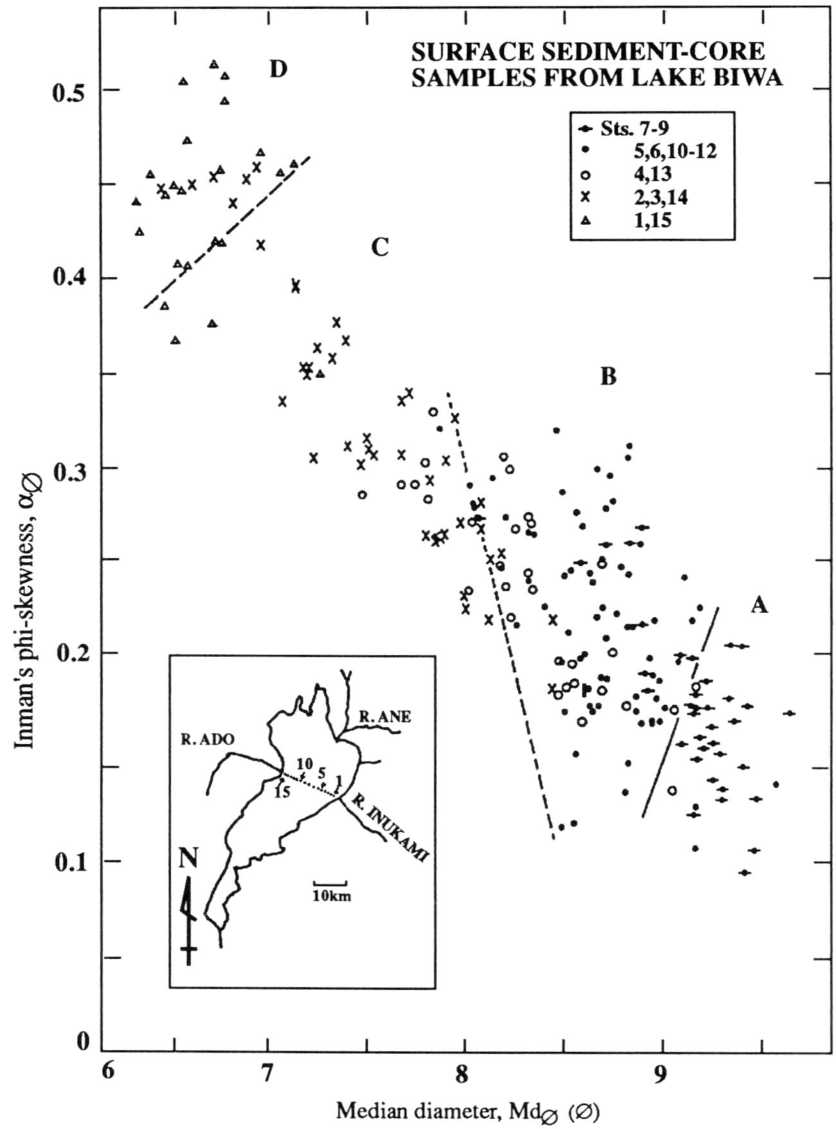


Figure 14.2. $Md_\phi$ - $\alpha_\phi$ diagram, separating the deposition area (domains A and B) from the shore area (domains C and D), using Krumbein's $\phi$ units where $\phi = -\log_2 d$, d being the grain size diameter (mm).

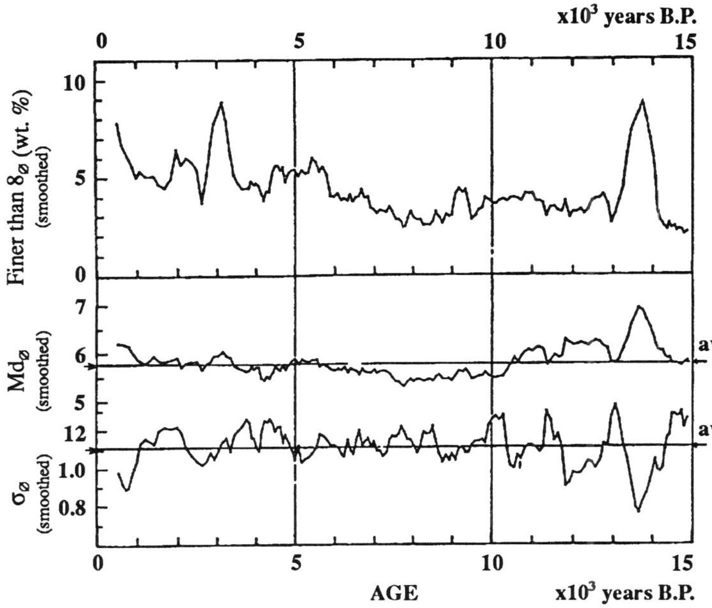


Figure 14.3. Smoothed variations in the grain size parameters: clay fractions (finer than $8_{\phi}$), phimedian diameter ($Md_{\phi}$) and phi-deviation measure ($\sigma_{\phi}$), in the 12 m core.

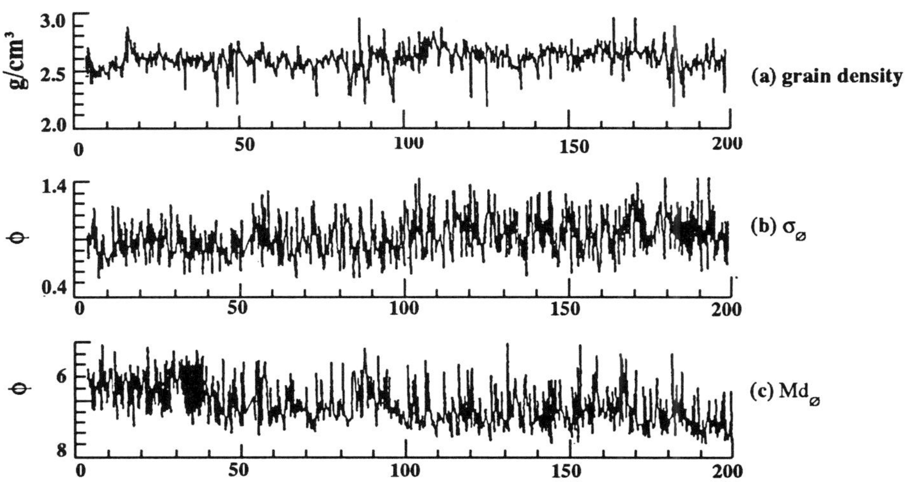


Figure 14.4. Variations in the grain size parameters in the 200 m core , (a) grain density, (b) $\sigma_{\phi}$ and (c) $Md_{\phi}$.

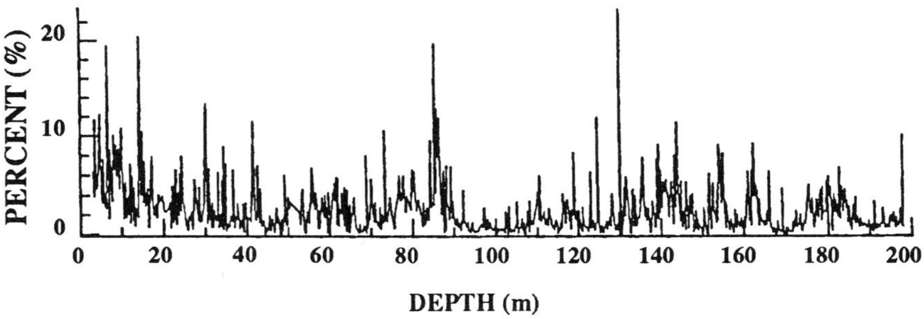


Figure 14.5. Paleoprecipitational change (the percentage of materials coarser than 44 μm) in the 200 m core.

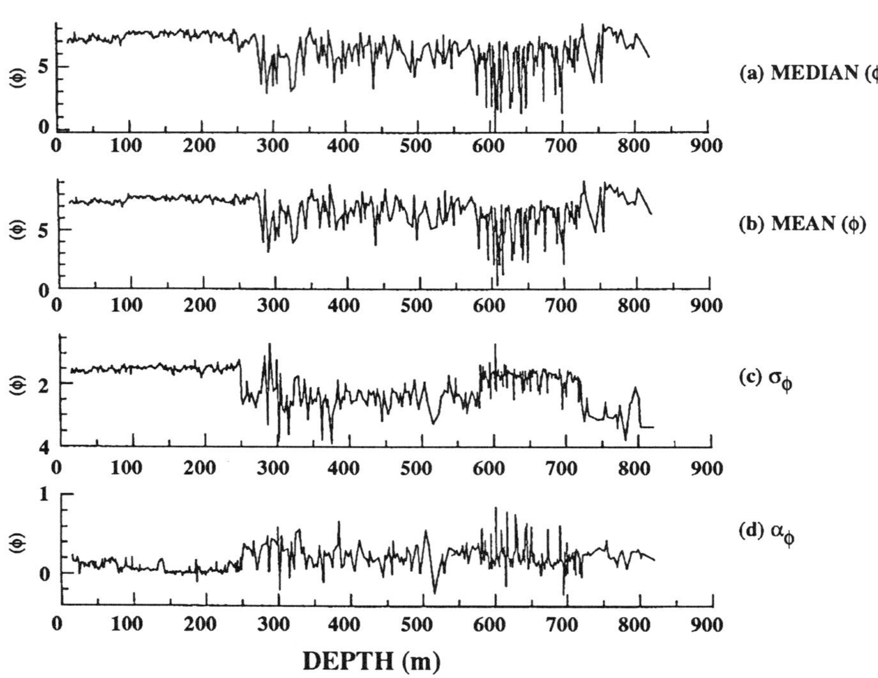


Figure 14.6. Variations in the grain size parameters in the 1,400 m core.

Short core sediments obtained from the area, belonging to domain A in Figure 14.2, consisted almost entirely of fine silt and clay. This deeper core, however, was composed of fractions coarser than 44 μm with comparatively large fluctuations in the grain size compared to the short cores. In periods of heavy precipitation, discharge from rivers increases and the density of river water becomes greater due to the large amount of suspended sediments. In such floods, turbidity currents, which may often also be low in temperature due to the effects of snow melting, are able to transport coarse sediments like sands directly to the central basin to sites usually reached only by very fine sediments. Assuming a constant geometry of the lake, the time variation in coarse particles might be regarded as a measure of changes in the sediment discharge in the past. Thus, the diagram of coarse fractions might contain some information on the variation in paleoprecipitation around Lake Biwa since 0.5 My B.P. (Figure 14.5).

The 1,400 m core was examined by analyzing 282 pieces of 20 cm long samples, taken out at intervals of 3 m, along the core down to 820 m. In grain size analysis, the X-ray extinction method was used instead of the photo-extinction method. Vertical variations in the bulk density, grain density, and grain size parameters are shown in Figure 14.6. The core was divided into 5 lithological groups. The uppermost group, the T-layer came from depths shallower than 250 m, consisted of very fine sediments lacking in broad fluctuations. The 2nd group, the S-layer extended down to 580 m, consisted of layers of sand alternating with gravels and sandy muds. The 3rd group, the R-layer down to 730 m in depth, consisted mainly of sediments similar to those in the T-layer. The 4th group, the Q-layer down to 820 m, consisted of layers of sands and gravels. Beneath the underlying group of gravel layers, the basement rocks (P-layer) were found.

The age of the lowest sub-layer (240 m deep layer) in the T-layer of the 1,400 m core which contained the whole section corresponding to the 200 m core, was estimated to be about 630 ky B.P. by Taishi *et al.* (1986). With the reference age of 450 ky B.P. at 180 m, they used the profile of bulk density measured by the fission-track dating method and a somewhat refined consolidation model where the sedimentation rate was assumed to be constant. The estimated rate of deposition was 33 mg cm$^{-2}$ y$^{-1}$, which is equivalent to the value of 0.3 mm y$^{-1}$ in the lower part of the T-layer. The variation in sand content in this layer is also considered to be due to paleoprecipitational change during the past 630 ky, in a manner similar to the case of the 200 m core (Figure 14.7).

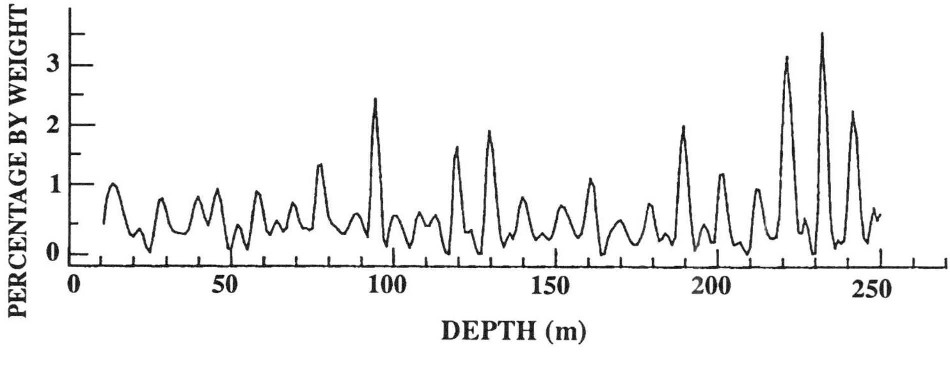


Figure 14.7. Variations in the sand content as a measure of precipitation, in the T-layer of the 1,400 m core.

## Periodicities Suggesting the Inference of Climatic Change from Grain Size Variations

The Milankovitch theory, which explains variations in the Pleistocene climate by changes in the earth's axis, precession and eccentricity of orbit, is one of the most significant theories concerning secular-climatic change (Milankovitch, 1941). The paleoclimatic change should result in variations in the erosional forces through changes of precipitation, temperature and wind systems. High precipitation rates enhance erosional forces, thereby increasing the amount of sediment transport and subsequent deposition of coarser grains on the lake bottom. Thus, variations in the grain size distribution of lake sediments may reflect paleoclimatic changes due to their effect on erosional processes. If the variation of solar insolation controlled by the earth's orbital parameters is important in forcing paleoclimatic change, as predicted by the Milankovitch theory, then the fundamental orbital periodicities should be reflected in the climatic record in the sediments.

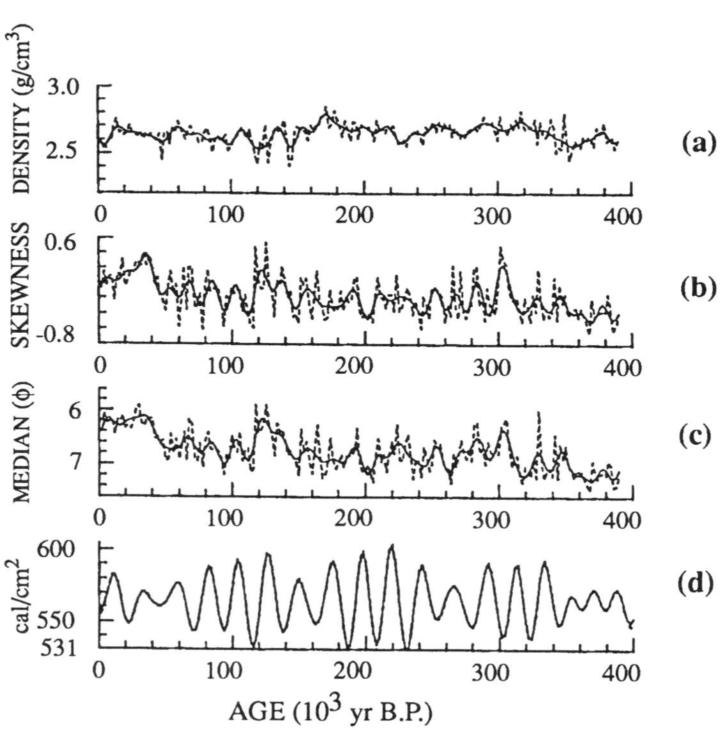


Figure 14.8. Comparison between insolation in the Northern Hemisphere in summer, (a) grain density, (b) skewness ($\alpha^2_\phi$), (c) median ($Md_\phi$) and (d) daily solar insolation averaged over the Northern Hemisphere calculated from Berger's formula, (1978). Age scale is based on the tentative one given by Kashiwaya et al., (1991).

The data sets of the grain size parameters from the 200 m core were arranged for statistical analysis in 2,000 year intervals. As seen in Figure 14.8 the smoothed fluctuations of the median and the 2nd phi skewness measure ($\alpha^2{}_\phi$) of the grain size distribution correspond closely with the fluctuations in the insolation. The $\alpha^2{}_\phi$ parameter appears to be a precise indicator of the modification of grain size distribution caused by the inflow of coarse grains.

The results of a spectral analysis of these data sets are shown in Figure 14.9. Three dominant periods, 18,000, 24,000 and 41,000 years predicted by the Milankovitch theory, are clearly detected in the grain size parameters (particularly $\alpha^2{}_\phi$ and also in the variation of boreal conifers by pollen analysis of the same core by Fuji (1978)). However, an important but weak period of 100,000 years seen in the insolation was not found in the 200 m core. Instead, additional and strong oscillations at 120,000 to 140,000 year and 60,000 to 70,000 year periods were indicated by the grain size, as well as by the pollen in the Lake Biwa sediments (Yamamoto *et al.*, 1985; Kashiwaya *et al.*, 1987; 1988). We can see similar periods in the grain size distribution to this T-layer in the 1,400 m core (Figure 14.10a), similar to those obtained in the 200 m core (Figure 14.9c), though those in the T-layer are a little larger.

## Change in Sedimentary Environments in Lake Biwa

Assuming that the $Md_\phi - \alpha_\phi$ relationship in the distribution pattern of the bottom sediments shown in Figure 14.2 holds for all the samples from Lake Biwa, then we can plot the data from the 1,400 m core on the same diagram to determine the domain that the present drilling site belonged to in the past. The resulting diagram is shown in Figure 14.11.

The T-layer is considered to have been deposited in environments similar to those at present, with respect to the lake geometry, because most of the sediments lie in the A and B domains. The R-layer (580 m to 730 m) seems to have been deposited in limnetic environments, not as deep as at present and fairly stable because all sediments lay in domain C. Unlike these two groups, the S-layer composed primarily of sands and gravels, the several intermittent narrow layers of muds lay in domains C or D and the rest remained outside the diagram. Therefore, it is supposed that the S-layer was deposited in non-limnetic environments.

In order to investigate in detail the changes in paleoenvironments indicated by the R-layer, grain size analysis was again carried out for 150 pieces of 5 cm long samples at 1 m intervals. Sieving with a Sonic Sifter was used for grains coarser than 125 μm, and the Laser Diffraction method (SALD 1000) was used for finer fractions (Shimadzu Co. information). Vertical variations in several parameters are illustrated in Figure 14.12. The R-layer contained many sandy layers and the parameters fluctuated fairly widely. The mean values of $Md_\phi$ and $\alpha_\phi$ were 5.29 and 0.22, respectively. The increase in $\alpha_\phi$ corresponds closely to the decreases in $Md_\phi$. The lake at that time might have been smaller, or at the drilling site, shallower than at present. Nevertheless, variations in the sand content (e.g. coarser than 125 μm) still suggest the paleoprecipitational change as in the T-layer, even if the curves were modified by temporary amplification or long-range trends due to some influences of crustal movement.

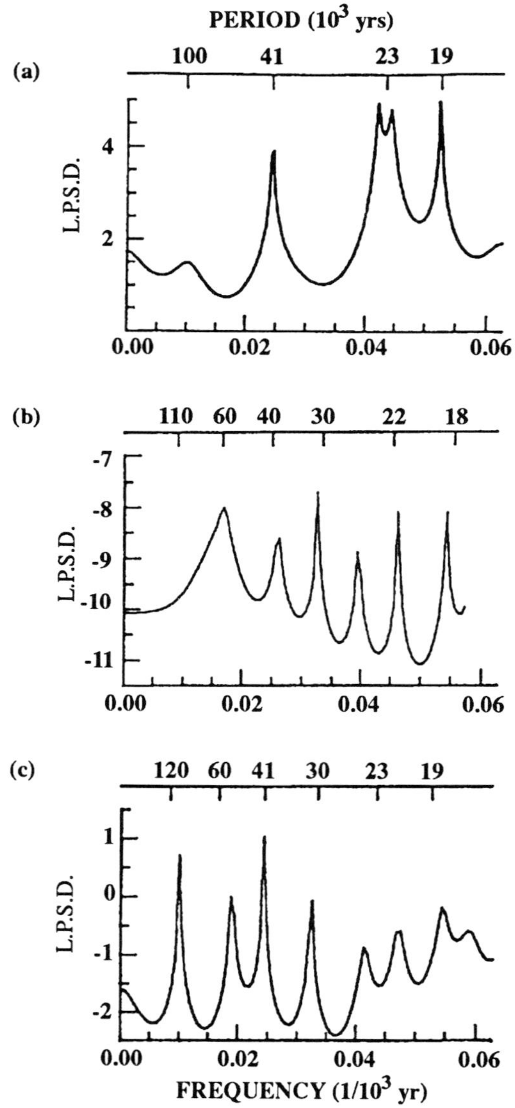


Figure 14.9. Log power spectral density for (a) insolation, (b) $\alpha^2_{\emptyset}$,(c) $Md_{\emptyset}$ . Data for (b) and (c) is from the 200 m core.

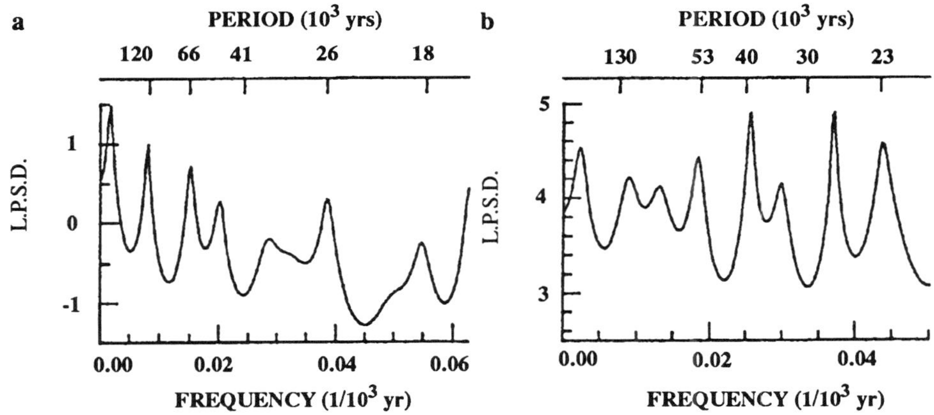


Figure 14.10. Comparison of dominant periods detected for (a) $Md_\phi$ in the T-layer and (b) sand content (coarser than $4_\phi$) in the R-layer of the 1,400 m core.

Bulk density values of 2.1-2.2 g $cm^{-3}$ suggest that the R-layer was fully consolidated. To discuss the grain size data statistically, the sedimentation rate was assumed to be constant throughout the R-layer and hence the thickness of the deposits was a linear function of time. Employing the constant rate of 0.3 mm $y^{-1}$ i.e. 3 m per 10,000 years, (actually the rate may be larger) obtained for the T-layer from the spectral analysis (Figure 14.10a), we found several dominant periods: ca. 23,000, 30,000, 40,000, 60,000 and 130,000 years. This set of periods seems to be in agreement with that in the T-layer (Figure 14.10a) and the 200 m core (Figure 14.9c).

According to the magnetostratigraphic study of the core, there exists a short normal polarity subzone which may be regarded as the Olduvai event at layers of 603-639 m, within the R-layer belonging to the Matuyama reversed polarity zone (Torii *et al.*, 1983). A fission-track date 2.0 (± 0.4) Ma given for the volcanic ash layer at 635 m in the layers by Yokoyama (1986) supports this inference. Age limits of the Olduvai event are usually regarded as 1.67 and 1.87 Ma. Assigning the ash layer at 635 m an age of 1.84 Ma by interpolation, we estimated that Lake Biwa originated around 2.16 Ma, and after the R-layer deposition had continued for 0.5 Ma, the environment of the lake underwent large changes at 1.66 Ma (Figure 14.12).

# Conclusion

On the basis of geographical distribution of sediment grain size, a $Md_\phi$ - $\alpha_\phi$ diagram of sedimentational zonation of the lake bottom was produced. From the results, a water depth of about 50 m divides the lake bottom into a deposition area and a shore area.

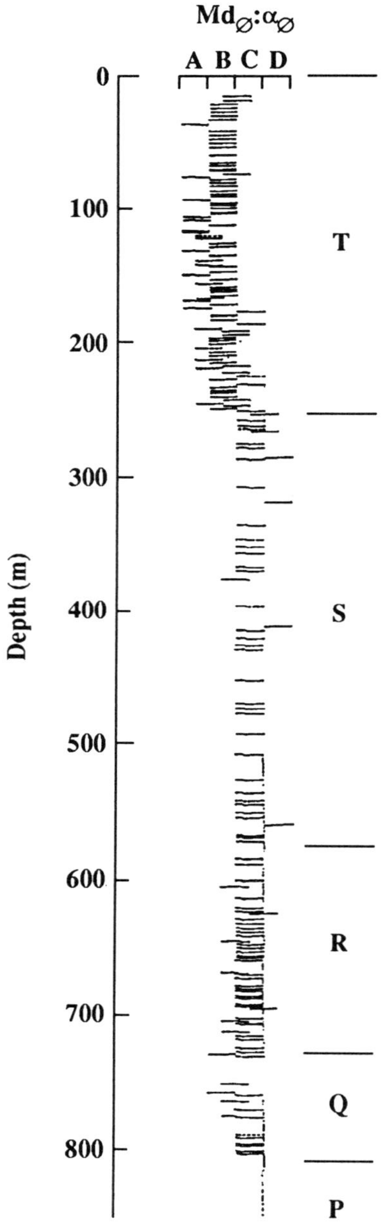


Figure 14.11. Vertical change of the domain occupied by sediments of the 1,400 m core on the $Md_{\varnothing}$ - $\alpha_{\varnothing}$ diagram shown in Figure 14.2.

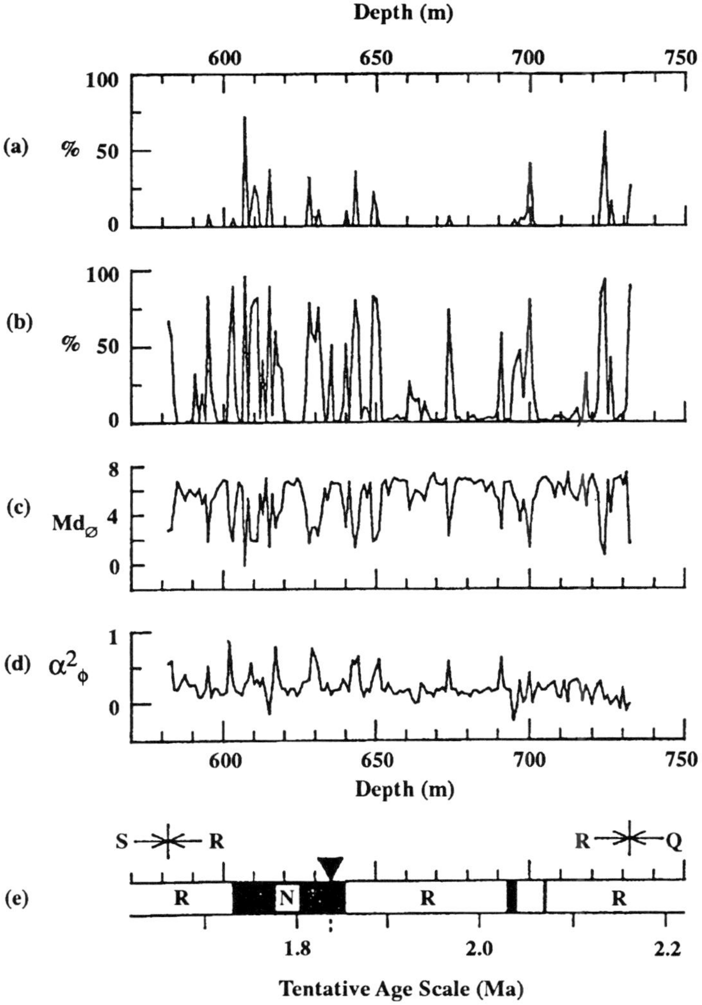


Figure 14.12. Vertical variations of grain size parameters in the R-layer of the 1,400 m core with tentative age scale. (a) sand content (coarser than 1$\varnothing$), (b) sand content (coarser than 3$\varnothing$), (c) median (Md$_{\varnothing}$), (d) skewness ($\alpha^2_{\varnothing}$) and (e) geomagnetic field.

Detailed grain size analysis of the 200 m core taken from the deposition area in the central part of Lake Biwa indicated variations in those contents coarser than 44 μm. In addition, paleoprecipitational changes since 500,000 years B.P. were estimated on the premise that coarse fractions like sand might have been brought in by riverine turbidity currents in periods of heavy precipitation (i.e. in pluvials).

From the spectral analysis of time series of grain size parameters for the 200 m core, dominant periods occurred at 19,000, 23,000, and 41,000 years. The Milankovitch theory proposes that such periods controlled climatic change. Furthermore, other dominant periods of 60,000 - 70,000 and 120,000 - 140,000 years were also detected. For the uppermost T-layer of the 1,400 m core, similar results were obtained.

The $Md_{\phi}$ - $\alpha_{\phi}$ diagram showed that the S-layer formed in non-limnetic environments and that the R-layer was deposited in a lake shallower than at present. The origin of Lake Biwa, implied by the R-layer and assuming that the sedimentation rate had a constant value similar to that in the T-layer, was estimated to have occurred at 2.16 Ma, and the duration of the lake era was 0.5 Ma before the lake underwent major changes.

# References

Berger, A. (1978) Long term variations of daily and quaternary climate changes. *J. Atmos. Sci.*, 35: 2362-2366.

Fuji, N., (1978) Palaeovegetational and palaeoclimatic changes around Lake Biwa, central Japan, during the past 100,000-170,000 years. *Paleolimnology of Lake Biwa and the Japanese Pleistocene*. 6: 235-262.

Horie, S. [Ed.], (1972-1983) *Paleolimnology of Lake Biwa and the Japanese Pleistocene*: 1-11.

Horie, S., O. Mitamura, S. Kanari, H. Miyake, A. Yamamoto and N. Fuji, (1971) Paleolimnological study on lacustrine sediments of Lake Biwa-ko. *Dis. Pre. Res. Inst. Ann., Kyoto Univ*. 14 B: 745-762 (In Japanese).

Kamitani, T., (1988) Grain size distribution of muds and their depositional process in Lake Biwa. *J. Res. Gr. Clas. Sed. Japan*, 5: 1-9 (in Japanese).

Kashiwaya, K., A. Yamamoto and K. Fukuyama, (1987) Time variations of erosional force and grain size in Pleistocene lake sediments. *Quat. Res*. 28: 61-68.

Kashiwaya, K., A. Yamamoto and K. Fukuyama, (1988) Statistical analysis of grain size distribution in Pleistocene sediments from Lake Biwa, Japan. *Quat. Res*. 30: 12-18.

Kashiwaya, K., K. Fukuyama and A. Yamamoto, (1991) Time variations in coarse materials from lake bottom sediments and secular paleoclimatic change. *Geophys. Res. Lett.* 18: 1245-1248.

Milankovitch M., (1941) Kanon der Erdbestrahlung und seine Anwendung auf das Eiszeitenproblem. Belgrad: Königl.serb.Akademie. 633pp. (Canon of insolation and the ice-age problem, English translation by the Israel Program for Scientific Translations, Jerusalem, 1969).

Morikawa, M., K. Kubo and Z. Sakaguchi, (1957) On the distribution of grain size of the sediments in Lake Biwa. *Bull. Shiga Univ*. 6: 39-46. (in Japanese)

Taishi, H., A. Yamamoto and S. Kanari, (1986) Age-scaling for the uppermost clayey layer in the 1,400 m core from Lake Biwa - A preliminary study. *Jpn. J. Limnol.* 47: 101-108.

Torii, M., H. Shibuya, A. Hayashida, I. Katsura, S. Yoshida, T. Tagami, Y. Otofuji and S. Sasajima, (1983) Paleomagnetic measurements of deep drilling core sample from Lake Biwa. *Rock Mag. Paleogeophys*. 10: 33-37.

Yamamoto, A., (1977) The structure of density and grain size variations seen near the 130 m layer in the 200 m long core sample from Lake Biwa. *Paleolimnology of Lake Biwa and the Japanese Pleistocene*. 5: 125-136.

Yamamoto, A., (1984) Grain size variation, pp.439-459. In: *Lake Biwa*, edited by S. Horie, Dordrecht: Dr. Junk Publ.

Yamamoto, A., M. Higashihara, T. Meguro, and H. Taishi, (1982) On properties of bottom sediments and sedimentation processes in Lake Biwa(II). *Paleolimnology of Lake Biwa and the Japanese Pleistocene*, 10: 59-71.

Yamamoto, A., S. Kanari and Y. Fukuo, (1973) Grain size analysis of sediments and their consolidation in a core from Lake Biwa. *Jpn. J. Limnol.* 34: 63-74. (in Japanese)

Yamamoto, A., Kashiwaya, K. and K. Fukuyama, (1985) Periodic variations of grain size in Pleistocene sediments in Lake Biwa and earth-orbital cycles. *Geophys. Res. Letter*, 12: 585-588.

Yokoyama, T., (1986) Tephrostratigraphy of sediment of actual Lake Biwa based on refractive indices of volcanic glasses in the deep drilled cores from the bottom of Lake Biwa, Japan. J. Geol. Soc. Jpan, 92: 653-661.

15

# Global Paleo-Hydrological Processes and Lake Biwa

K. Kashiwaya and K. Fukuyama

## Introduction

Hydrological conditions often affect the earth's surface processes in areas with humid climates, such as East Asia. Specifically, heavy rainfall erodes the earth's surface and changes landforms. Part of this eroded material is transported to seas and lakes, and sediments in closed or semi-closed bains clearly reflect such erosional conditions. Grain size variations of the sediments can reveal information on the magnitude of the erosional force; large grains correspond to a large erosional force and small grains to a small force, all other conditions being equal. Therefore, it should be possible to trace temporal changes in the hydrological conditions by analyzing the grain size distributions of lake sediments.

The Milankovitch theory (1941) on the causes of the variations in the Pleistocene climate is of great value to research on erosional forces (Kashiwaya *et al.*, 1987; 1988) as well as in determining the causal relation between insolation and secular change in the climate (e.g. Hays *et al*., 1976; Berger *et al*., 1984). Milankovitch parameters have been found in the grain size variations of Lake Biwa sediments, as stated in the preceding paper (also, cf., Yamamoto *et al*., 1984; 1985). Not all the dominant periods found were clearly predicted by the Milankovitch theory (Kashiwaya *et al.*, 1988; 1991b). The periods are however, very important for investigating the secular change in the hydrological conditions. In this paper, we shall first discuss solar insolation and some secular climatic changes from various areas (cf. Kashiwaya *et al.*, 1990; Fukuyama *et al*., 1991), and then investigate the grain size variations from Lake Biwa in view of paleohydrological change and compare it with the climatic changes.

Physical Processes in a Large Lake: Lake Biwa, Japan
Coastal and Estuarine Studies, Volume 48, Pages 169–180

## Insolation and Sea Bottom Cores

In the Milankovitch theory (Milankovitch, 1941), solar climate is clearly expressed by secular change in the solar insolation. Evidence for changes in insolation was found in several deep sea cores (cf. Hays *et al.*, 1976; Janecek and Rea, 1984). However, secular changes in the climatic system cannot always be explained by this theory (Broecker and Denton, 1989). For example, the largest oscillation with a 100,000-year period has not been adequately explained by this theory since this period has only a very small amplitude in the eccentricitical parameter. Various models have tried to explain how the extremely small amplitude became the largest fluctuation (e.g., Lindzen, 1986). There remain, however, chronological and methodological problems in discussing the period

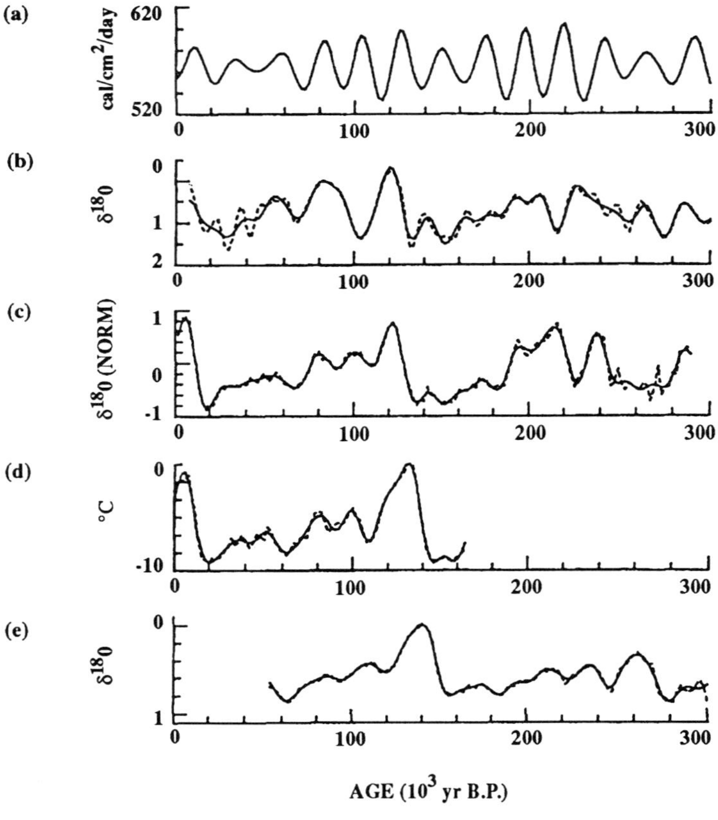


Figure 15.1. Time variations in data sets discussed in this paper. (a) Insolation curve average over the Northern Hemisphere (Berger, 1978), (b) Marine $\delta^{18}O$ from RC11-210 (Chuey *et al.*, 1987) and (c) $\delta^{18}O$ record of the RC11-120 (Martinson *et al.*, 1987). (d) Antarctic surface temperature change (Jouzel *et al.*, 1987; Lorius *et al.*, 1988). (e) Continental $\delta^{18}O$ (Winograd *et al.*, 1988). Solid lines indicate low-pass-filtered curves (filter removing components of periods less than 15,000 years) and dotted lines show the original data.

thoroughly. Here, we applied the same statistical method (spectral analysis, harmonic analysis and filter analysis) as was used to analyze Lake Biwa data (Yamamoto *et al.*, 1984; 1985, Kashiwaya *et al.*, 1987; 1988; 1990, Fukuyama *et al.*, 1991) to analyze some marine data in order to compare the periods with those obtained from the Lake Biwa record. The data sets discussed in this paper are shown in Figure 15.1 and 15.2.

An insolation curve averaged over the Northern Hemisphere for summer was used in our analysis based on Berger's formula (1978) (Figure 15.1a). Recently, Berger and Loutre (1989) reported a new version for the calculation of insolation, but there is no modification in the range of the past 300 - 500 millennia as is discussed in this paper.

Two marine $\delta^{18}O$ data sets were used here: one from the RC11-210 core from the equatorial central Pacific, analyzed by Chuey *et al.* (1987) and the other from the RC11-120 core from the Indian Ocean, analyzed by Martinson *et al.* (1987) (Figures 15.1b and 15.1c respectively). The Pleistocene $\delta^{18}O$ fluctuations in deep-sea cores are said to reflect changes in the oceanic oxygen isotope composition, caused primarily by ice-sheet growth and decay (Shackleton, 1967).

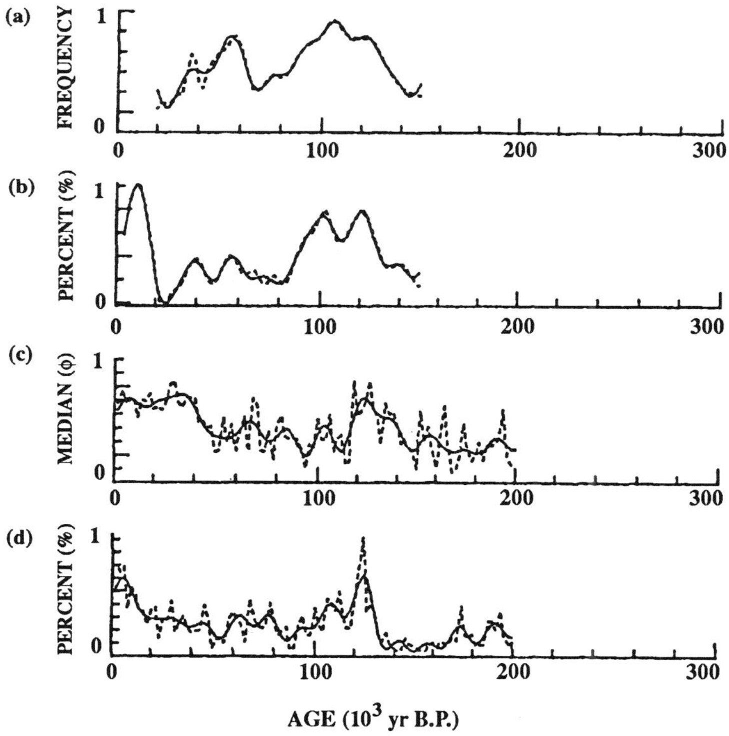


Figure 15.2. (a) speleothem growth frequency in Britain (Gordon *et al.*, 1989) and (b) speleothem growth frequency compiled by Hennig *et al.*, 1983, (c) the median of the grain size distribution and (d) the percentage of materials coarser than 4.5ϕ in the sediment from Lake Biwa. Solid curves indicate low-pass-filtered curves and dotted lines mean the original data.

## Statistical Analysis of Insolation

The results of spectral analysis (Barrodale and Ericksson, 1980a; 1980b) for the insolation curve averaged over the Northern Hemisphere revealed several dominant periods; some in addition to the four famous periods, 100,000, 41,000, 23,000 and 19,000 years (cf. Figure 14.9 in Chapter 14). Figure 15.3 depicts filtered insolation time series where the size of individual orbital components is clearly expressed. As can be seen from these figures, the amplitude of the eccentricity is extremely small, compared with those of other periods from the precessional and obliquitical parameters. To compare the sizes of the parameters, harmonic analysis was employed (Table 15.1). The results show that the amplitude of the 100,000-year period is about a sixth to a tenth of the obliquitical parameter and a tenth to a fiftieth of the precessional parameters. We found larger amplitudes around 30,000 and 60,000, though they were much smaller than those in precessional and obliquitical parameters. These results lead to the belief that the eccentricitical parameter contributes very little to the insolation.

TABLE 15.1. Amplitudes of prevailing periods in the isolation averaged over the Northern Hemisphere.

| Period (ky) | Amplitude |
|---|---|
| 113.91 | 0.29 |
| 91.33 | 0.21 |
| 56.95 | 0.35 |
| 40.97 | 3.86 |
| 28.85 | 0.61 |
| 23.56 | 15.10 |
| 18.89 | 9.71 |

## Statistical Analysis of Marine $\delta^{18}O$

The results of Chuey *et al*. (1987) and Martinson *et al*. (1987) show that there exist three or four obliquitical (41,000 years) and precessional parameters (18,000-19,000 years and 22,000-24,000 years) and the periods around 100,000 years (Figure 15.4). There are two periods around 100,000-year period: 120,000 and 60,000 years for the RC11-210 data and 110,000 years and 70,000 years for RC11-120 data. Though low frequency oscillations are not always clearly resolved by spectral analysis due to the finite length of records, the results of harmonic analysis for two datasets (Table 15.2) indicate that there are some periods around 100,000-year and the amplitude of the 100,000- to 120,00-year period is largest among them. These results indicate that the longest period of 100,000 years established by Hays *et al.* (1976) appears not to be a simple 100,000-year period, but rather a combination of periods in the range of 110,000 to 130,000 years and weaker periods of 60,000 to 70,000 years.

TABLE 15.2. Amplitudes and phases from harmonic analysis of prevailing periods for the marine $\delta^{18}O$ record of the (a) RC11-210 core and (b) RC11-120 core.

| (a) | | | (b) | | |
|---|---|---|---|---|---|
| Period (ky) | Amplitude | Phase (ky) | Period (ky) | Amplitude | Phase (ky) |
| 122.9 | 0.261 | 90.9 | 111.3 | 0.422 | 102.0 |
| 61.0 | 0.153 | 1.1 | 74.0 | 0.116 | 51.7 |
| 37.5 | 0.241 | 9.0 | 58.2 | 0.088 | 56.4 |
| 28.0 | 0.191 | 5.3 | 40.0 | 0.216 | 3.5 |
| 23.4 | 0.094 | 2.4 | 29.2 | 0.085 | 5.5 |
| 21.0 | 0.162 | 15.6 | 23.3 | 0.227 | 6.5 |
| | | | 18.9 | 0.110 | 7.6 |

TABLE 15.3. Amplitudes and phases of prevailing periods for (a) the Vostok ice core and for (b) the $\delta^{18}O$ record from the Great Basin from harmonic analysis.

| (a) | | | (b) | | |
|---|---|---|---|---|---|
| Period (ky) | Amplitude | Phase (ky) | Period (ky) | Amplitude | Phase (ky) |
| 124.9 | 2.24 | 114.0 | 115.9 | 0.175 | 10.4 |
| 62.7 | 1.43 | 1.5 | 60.0 | 0.095 | 19.5 |
| 42.5 | 2.15 | 1.7 | 44.9 | 0.067 | 1.8 |
| 31.6 | 0.87 | 5.4 | 30.5 | 0.103 | 17.9 |
| 24.7 | 0.83 | 4.6 | 26.0 | 0.043 | 6.2 |
| 18.2 | 0.59 | 8.1 | 18.7 | 0.021 | 12.0 |

## Hydrological Secular Variation from Continental Sediments

### *Vostok Core and Great Basin Core*

Datasets from the Vostock and Great Basin Cores were of great use not only in discussing paleoclimates but also in reconstructing paleohydrological conditions. The Vostok ice core extracted in 1980-1982 at Vostok station in the central part of East

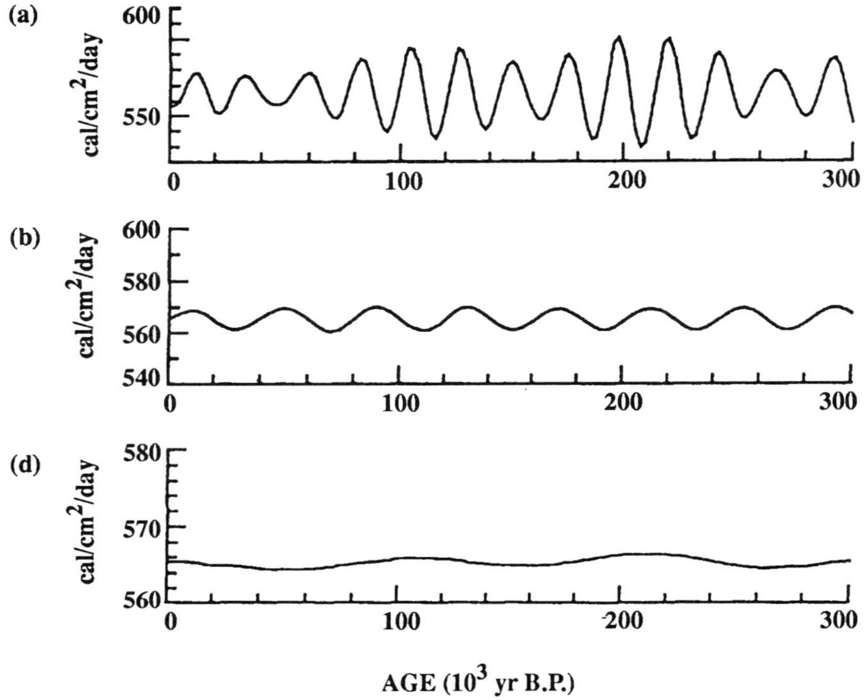


Figure 15.3. Insolation curve averaged over the Northern Hemisphere; (a) 30,000- to 15,000-year band-pass-filtered curve, (b) 50,000- to 15,000-year band-pass-filtered curve and (c) 80,000-year low-pass-filtered curve.

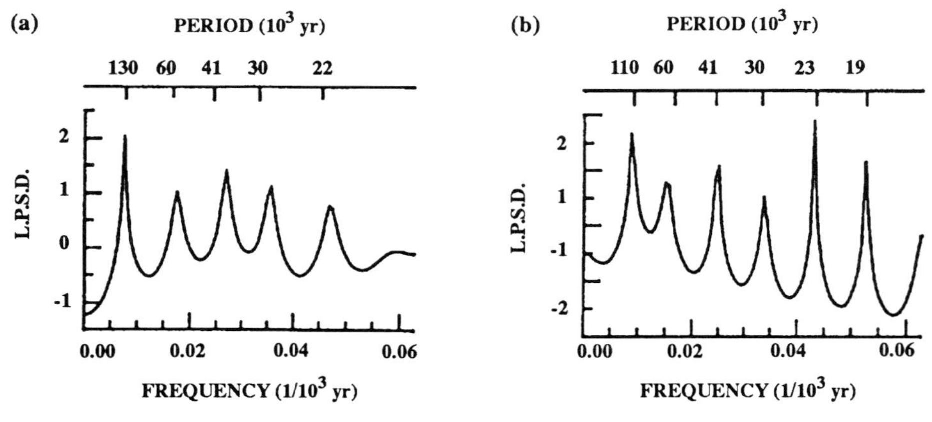


Figure 15.4. Log power spectral density for the marine $\delta^{18}O$ record of (a) the RC11-210 and (b) RC11-120.

Antarctica (Lorius *et al.*, 1985) revealed a great deal of information on paleoclimatic conditions. Here we discuss a dataset on the temporal change in surface temperature (Figure 15.1d) deduced from an isotope record (Jouzel *et al.*, 1987; Lorius *et al.*, 1988) since we have used a similar technique above. It was also discovered that a change in the $CO_2$ concentration was closely related to that of temperature (Barnola *et al.*, 1987) and it is presumed that this change had some influence on the hydrological conditions.

The $\delta\ ^{18}O$ record (Figure 15.1e) from a calcite vein in the Great Basin showed changes in the paleohydrological conditions in the area (Winograd *et al.*, 1988). In this case, the calcite was deposited from ground water moving through an open fault zone in Southern Nevada. As pointed out by Winograd *et al.* (1988), there are some phase lags between the temporal change in the data obtained here and changes found in other $\delta\ ^{18}O$ datasets, but there are no inconsistencies concerning the trend and periodicity of the data.

Figure 15.5 shows the results of the spectral analysis for the two datasets. We again found no period of exactly 100,000 years, though we found obliquitical (41,000 years) and precessional parameters (18,000-19,000 years and 24,000-25,000 years) in both datasets. Instead, there are 116,000 and 60,000-year periods around the 100,000-year period in the Great Basin data and 130,000 and 60,000-year periods in the Vostok data. Thus these results also indicate that the largest 100,000-year period reported by Hays *et al.* (1976) may be a combination of 110,000- to 130,000-year and 60,000- to 70,000-year periods.

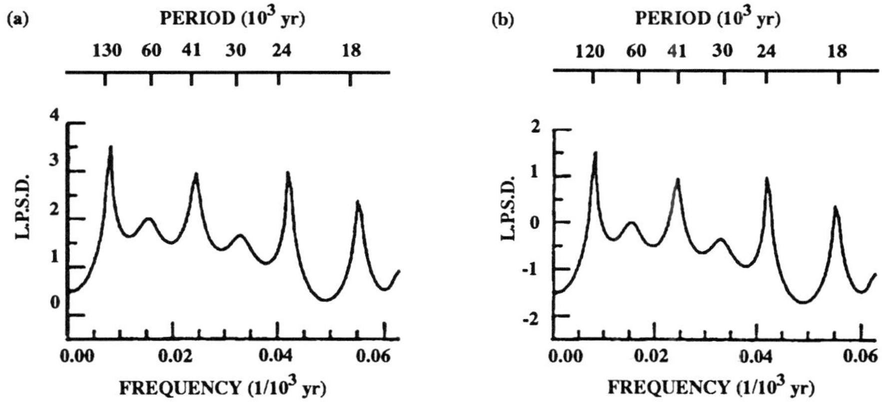


Figure 15.5. Log power spectral density for the Antarctic surface temperature change derived from (a) the Vostok ice core and for (b) the continental $\delta\ ^{18}O$ record from the Great Basin.

## *Speleothem Data*

Speleothem growth is controlled by several climatic and environmental factors, of which water availability, temperature and vegetation cover are the most significant (Atkinson *et al.*, 1978; 1986; Hennig *et al.*, 1983). Consequently, data on speleothem growth frequency from the past may contain useful information on paleohydrological conditions. Recently, Kashiwaya *et al.* (1991a) for example, found that there are some close correlations between speleothem abundance and solar insolation.

Two datasets are discussed (Figures 15.2a and 15.2b); one for speleothem growth frequency in Britain performed by Gordon *et al.* (1989) and another by Hennig *et al.* (1983) who compiled speleothem data for most of the Northern Hemisphere. The results of spectral analysis are presented in Figure 15.6, where two clear Milankovitch

parameters are present in the dataset from Britain: obliquitical (40,000-year period) and precessional parameters (20,000-year period). These parameters are not as clear as in the dataset from the Northern Hemisphere, though they are present. Two longer periods of 60,000 years and 120,000 to 130,000 years can be seen in both the datasets as would be expected if the datasets were connected with global climatic changes.

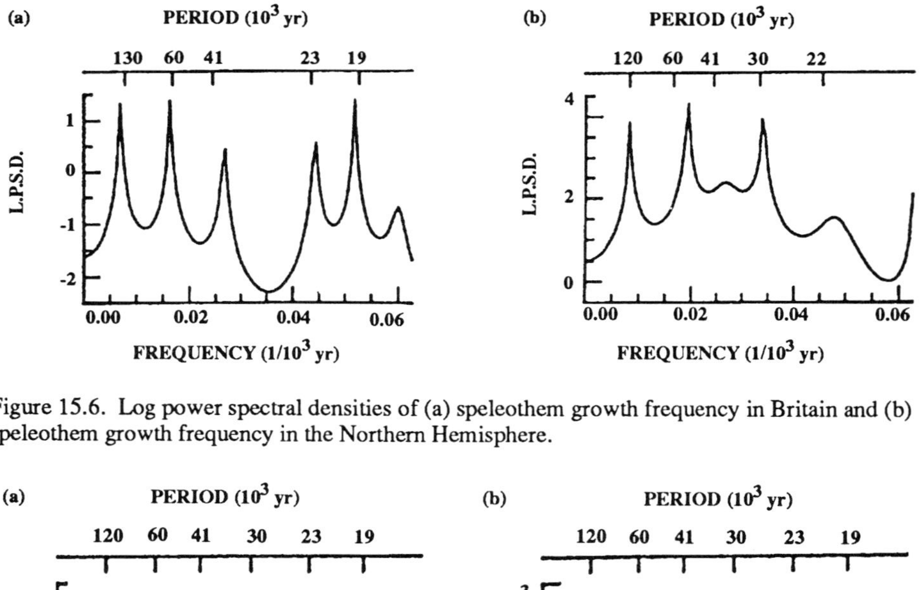


Figure 15.6. Log power spectral densities of (a) speleothem growth frequency in Britain and (b) speleothem growth frequency in the Northern Hemisphere.

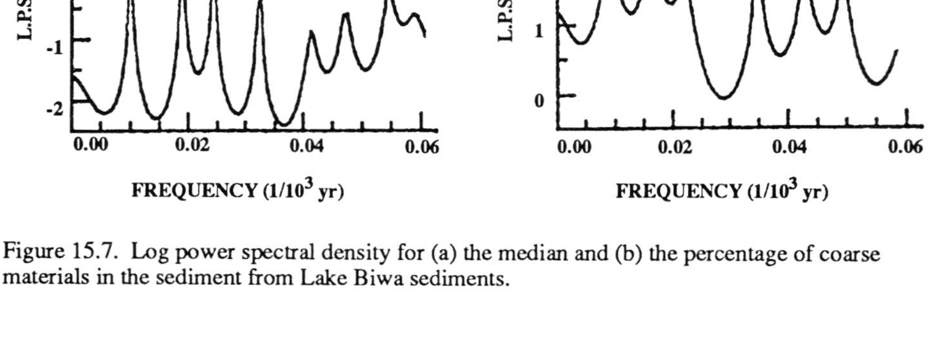


Figure 15.7. Log power spectral density for (a) the median and (b) the percentage of coarse materials in the sediment from Lake Biwa sediments.

# Paleohydrological Processes and Lake Biwa Sediments

As presumed in the above sections, lengthy cores sampled from Lake Biwa have revealed a large amount of valuable information on the paleoenvironment of the Japanese Pleistocene (e.g. Horie, 1984). Because of successive accumulation and rapid deposition rates, variations in the components of the Lake Biwa sediments have been very useful for reconstructing changes in surrounding environmental conditions during the past 500,000 years around the lake. Some components may include information on secular landform changes due to various erosional forces which were related to paleohydrological conditions (Kashiwaya, 1986, 1987; Kashiwaya *et al.*, 1987). Grain size distributions in bottom sediments, for example, reflect some hydrological conditions; coarser grains indicating strong inflows which were able to transport larger particles from the surrounding areas into the central part of the lake, while finer grains suggest weak inflows. Furthermore, the size of the inflows depends on the amount of precipitation and area of the lake.

Some parameters on grain size data from a 200-m long core are discussed in the preceding paper (Inman' parameters, graphical measure) and elsewhere (Yamamoto, 1984; Yamamoto *et al.*, 1984; 1985, Kashiwaya *et al.*, 1987; 1988; 1990). The median value and the percentage of materials coarser than 4.5 ø in the sediment of the upper 200,000 years of the core are shown in Figures 15.2c and 2d, and their power spectral density in Figure 15.7. Both sets of grain size data have about 140,000-, 70,000-, 40,000-, and 20,000-year periods, as previously reported for the whole data (Kashiwaya *et al.*, 1988). Of these periods, the 40,000- and 20,000-year periods can be predicted directly from the Milankovitch theory, but the 140,000- and 70,000-year periods cannot. The 140,000-year period may refer to a 120,000-year period of eccentricity, but it is not evident from this analysis alone. However, most secular changes in the various datasets discussed above have two periods of around 100,000 years. Therefore, the existence of these two periods may indicate that the grain size data from Lake Biwa were synchronized with global climatic changes through some changes in hydrological conditions.

The median value of the datasets also shows that coarser particles were deposited during periods of strong insolation, whereas finer particles were deposited during those periods of weak insolation. In general, there seems to be no distinct discrepancy in the median data between glacial periods and interglacial periods. Of course, there remain some chronological problems such as possible phase lags between the age scale employed here and the absolute age scale. On the other hand, we can see a clear difference between the last glacial period and the last interglacial period in the temperature datasets discussed. This may indicate that the change in the average values of the grain size distribution responded slowly to climatic change while the change in temperature is a climatic change in itself.

# Conclusion

We conclude that distinct periods of about 20,000 and 40,000 years exist in all the datasets and that they correspond to the periods for the obliquitical and precessional parameters. The 70,000-year period, found in both the grain size variation of the Lake Biwa sediments and in the RC11-230 core, was not seen in the insolation data, although there is a weak 60,000-year period in the insolation data. The 60,000-year period was present in all the datasets of the $\delta^{18}O$ and speleothem frequency and the 70,000-year period may be related to the 60,000-year period. The 110,000- to 120,000-year periods were present in the $\delta^{18}O$ of the RC11-120 core, Great Basin Data and speleothem frequency of the northern Hemisphere and most likely correspond to a 100,000- or 120,000-year period in the eccentricitical parameters (Table 15.1). The 130,000- to 140,000-year periods found in the variations of the RC11-230 core, Vostok data and grain size distribution from Lake Biwa may refer to the 120,000-year period in the eccentricitical parameters. However, the periods are so weak that they cannot be easily detected in the insolation curve. Therefore, it is not yet clear that they are based on eccentricitical parameters and they may be dependent upon other orbital parameters.

It is clear, however, that the longer periods characteristic of the grain size variations in the Lake Biwa records are present in all the datasets described in this chapter, though there are slight differences in the magnitudes and phases. This suggests that the grain size variations of Lake Biwa, which are considered to reflect paleohydrological conditions of the surrounding areas, are closely related to global climatic changes, though they may not have been directly influenced by continental ice sheets and eustatic sea level changes.

# References


Atkinson, T. C., R. S. Harmon, P. L. Smart and A. C. Waltham, (1978) Paleoclimatic and geomorphic implications of 230Th/234U dates on speleothems from Britain. *Nature* 272: 24-28.

Atkinson, T. C., T. J. Lawson, P. L. Smart, R. S. Harmon and J. W. Hess, (1986) New data on speleothem deposition and paleoclimate in Britain over the last forty thousand years. *J. Quart. Res*. 1: 67-72.

Barnola J. M., Y.S. Korotkevitch, D. Raynaud and C. Lorius, (1987) Vostok ice core provides 160,000-year record of atmospheric $CO_2$. *Nature* 329: 408-414.

Barrodale I. and R. E. Ericksson, (1980a) Algorithm for least-square linear prediction and maximum entropy spectral analysis, part 1. Theory. *Geophysics* 45: 420-432.

Barrodale I. and R. E. Ericksson, (1980b) Algorithm for least-square linear prediction and maximum entropy spectral analysis, part 2. Programme. *Geophysics* 45: 433-446.

Berger A., (1978) Long-term variations of daily and Quaternary climatic changes. *J. Atmos. Sci.* 35: 2362-2366.

Berger A. and M. F. Loutre, (1989) A new astronomical solution for the Neogene climate. Terra abstracts 1: 16 (Abstract of the *3rd I.C.P. held in Cambridge*). Oxford: Blackwell Sci. Pub. 93 pp.

Berger A., J. Imbrie, G. Kukla and B. Saltzman (eds), (1984) *Milankovitch and Climate - understanding the response to astronomical forcing*. Dordrecht: D. Reidel. 510 pp.

Broecker, W. S. and G. H. Denton, (1989) The role of ocean-atmospheric reorganization in glacial cycles. *Geochemica et Cosmochima Acta* 53: 2465-2501.

Chuey J. M., D. K. Rea, and N. G. Pisias, (1987) Late Pleistocene paleoclimatology of the central equatorial Pacific: A quantitative record of eolian and carbonate deposition. *Quat. Res*. 28: 323-339.

Fukuyama, K., K. Kashiwaya, K. and A. Yamamoto, (1991) Quasi-periodic oscillations in plate Pleistocene climatic change. *J. Meteor. Soc. Japan*. to be submitted.

Gordon, D., P. L. Smart, D. C. Ford, J. N. Andrews, T. C. Atkinson, P. J. Rowe and N. S. J. Christopher, (1989) Dating of late Pleistocene interglacial and interstadial periods in the United Kingdom from speleothem growth frequency. *Quat. Res* 31: 14-26.

Hays J. D., J. Imbrie, and N. J. Shackleton, (1976) Variations in the Earth's orbit: pacemaker of the ice ages. *Science* 194: 1121-1132.

Hennig, G.J., R. Grun, and K. Brunnacker, (1983) Speleothem, travertines, and paleoclimates. *Quat. Rese*. 20: 1-29.

Horie S. [Ed.], (1984) *Lake Biwa*. Dordrecht: Junk. 654 pp.

Janecek, T. R. and D. K. Rea, (1984) Pleistocene fluctuations in Northern Hemisphere tradewinds and westerlies, pp.331-347. In *Milankovitch and Climate - understanding the response to astronomical forcing*, edited by A. Berger, J. Imbrie, G. Kukla and B. Saltzman, Dordrecht: D. Reidel. 510pp.

Jouzel J., C. Lorius, J. R. Petit, C. Genthon, N. I. Barkov, V. M. Kotlyakov and V. M. Petrov, (1987) Vostok ice core: a continuous isotope temperature record over the last climatic cycle, (160,000 years). *Nature* 329: 403-408.

Kashiwaya, K., (1986) A mathematical model of the erosional process of a Mountain. Transactions, *J. Geomorph. Union* 7: 69-77.

Kashiwaya, K., (1987) Theoretical investigation of the time variation of drainage density. *Earth Surface Processes and Landforms* 12: 39-46.

Kashiwaya, K., T. C. Atkinson and L. S. Peter, (1991a) Periodic variations in late Pleistocene speleothem abundance in Britain. *Quat. Res.* 32.

Kashiwaya, K., K. Fukuyoma and A. Yamamoto, (1991b) Time variations in coarse materials from lake bottom sediment to secular paleoclimatic change. *Geophys. Res. Lett.* 18: 1245-1248.

Kashiwaya, K., A. Yamamoto and K. Fukuyama, (1987) Time variations of erosional force and grain size in Pleistocene lake sediments. *Quat. Res*. 28: 61-68.

Kashiwaya, K., A. Yamamoto and K. Fukuyama, (1988) Statistical analysis of grain size distribution in Pleistocene sediments from Lake Biwa, Japan. *Quat. Res*. 30: 61-68.

Kashiwaya, K., K. Fukuyama and A. Yamamoto, (1990) Erosional environment and grain size variation in Pleistocene lake sediments. *L'Anthropologie* 94.

Lindzen, R.S., (1986). A simple model for 100k-year oscillations in glaciation. *J. Atmos. Sci*. 43: 986-996.

Lindzen, R. S., and A. Y. Hou, (1988) Hadley circulations for zonally averaged heating centered off the equator. *J. Atmos. Sci*. 45: 2416-2427.

Lorius, C., J. Jouzel, C. Ritz, L. Merlivat, N. I. Barkov, Y. S. Korotkevich and V. M. Kotlyakov, (1985) A 150,000 year climatic record from Antarctic ice. *Nature* 316: 591-596.

Lorius, C., N. I. Barkov, J. Jouzel, Y. S. Korotkevich, V. M. Kotlyakov and D. Raynaud, (1988) Antarctic ice core: $CO_2$ and climatic change over the last climatic cycle. *Eos Trans. AGU*. 69: 681-684.

Martinson, D. G., N. G. Pisias, J. D. Hays, J. Imbrie, T. C. Moore Jr. and N. J. Shackleton, (1987) Age dating and orbital theory of the ice ages; development of a high-resolution 0 to 300,000-year chronostratigraphy. *Quat. Res*. 27: 1-29.

Milankovitch, M., (1941) Kanon der Erdbestrahlung und seine Andwendung auf das Eiszeitenproblem. Belgrad: Königlich Serbische Akademie. 633 pp. (Canon of insolation and the ice-age problem, English translation by the Israel Program for scientific translations, Jerusalem, 1969).

Shackleton, N. J., (1967) Oxygen isotope analysis and Pleistocene temperatures reassessed. *Nature* 218: 15-17.

Winograd, I, B. J. Szabo, T. B. Coplen, and A. C. Riggs, (1988) A 250,000-year climatic record from Great Basin vein calcite: Implications for Milankovitch theory. *Science* 242: 1275-1280.

Yamamoto, A., (1984) Grain size variation, pp.439-459. In *Lake Biwa.*, edited by S. Horie, Dordrecht: Junk. 645 pp.

Yamamoto, A, K. Kashiwaya and K. Fukuyama, (1984) Periodic variations of grain size in the 200 meter core sample from Lake Biwa. Trans. *J. Geomorph. Union* 5: 345-352.

Yamamoto, A, K. Kashiwaya and K. Fukuyama, (1985) Periodic variations of grain size in Pleistocene sediments in Lake Biwa and earth-orbital cycles. *Geophys. Res. Lett*. 12: 585-588.

# 16

# Thermal Infrared Mapping

Y. Itakura

## Introduction

Infrared remote sensing has many useful applications in meteorological and environmental studies in areas such as resource investigation, pollution surveillances and the location of fish. Two main techniques are available : spectral radiometry and thermal infrared mapping. Thermal infrared imagery of the ocean surface obtained from the satellite-borne sensors provides data on tidal currents and storms for meteorology, environmental and fishery sciences (McClain, 1985; Tameishi, 1985; Takayama and Maeda, 1990). In addition to the satellite's sensors, airborne sensors and sensors positioned on high buildings or mountain tops can also provide useful data on surface temperature distributions in lakes. Because the water density is primarily determined by the temperature it is possible to deduce the water movements in freshwater lakes, such as Lake Biwa, from the spatial and temporal variability of the water temperature. The movement of water in lakes is important not only in its own right, but also for the chemical and biological properties of the lake.

The sensors for thermal infrared mapping of the surface temperature of lakes fall into three categories.

(1) Satellite-borne sensors have the advantage of wide observation area, but suffer from problems due to disturbance of the observations by clouds and low spatial resolution on the surface.

(2) Airborne- and balloon-borne sensors have high spatial resolution and a wide observation area, but it is impossible to maintain continuous observation at a point and they are expensive to operate.

Physical Processes in a Large Lake: Lake Biwa, Japan
Coastal and Estuarine Studies, Volume 48, Pages 181–193

(3) Sensors positioned on high buildings or mountain tops enable continuous observation at a point and high spatial resolution, but have a slanted path of observation, suffer from interference from background radiation and have a narrow observation area.

The selection of an optimum sensor carrier depends on the practical objects of the research project as well as the observation situations.

## Temperature Estimation of Water Surface

### *Principle of Estimation*

A typical example of infrared remote sensing of the surface temperature of lake water is depicted in Figure 16.1. In this situation, the sensor is aimed towards a spot on the lake surface and measures the radiant flux emerging from the earth's atmosphere. The radiant flux $\Phi$ $(\phi,\psi)$ received, corresponding to the pixel of the thermal infrared mapper at a position $(\phi,\psi)$ contains: (1) Thermal radiation of the surface, (2) background radiation reflected at the surface, (3) thermal radiation of the atmosphere along the optical path of the sensor, and (4) background radiation scattered in the atmosphere along the optical path. The function $\Phi$ $(\phi,\psi)$ is given by the following equation:

$$\Phi(\phi,\psi) = (\pi/4)(A_d D^2/f^2)\int \tau_0(\lambda)\{[\varepsilon_\sigma(\lambda,\phi,\psi)\,L(\lambda,T_s) + (1-\varepsilon_\sigma(\lambda,\phi,\psi))\,L(\lambda,T_b)]\,\tau_a(\lambda,\phi,\psi) + (1-\tau_a(\lambda,\phi,\psi))\,L(\lambda,T_a)\}\,d\lambda, \quad (16.1)$$

where

| | |
|---|---|
| $f$ | is the focal length of the optics, |
| $D$ | is the aperture diameter of the optics, |
| $A_d$ | is the area of the detector, |
| $\tau_0(\lambda)$ | is the spectral transmittance of the optics containing the spectral response of the detector, |
| $\varepsilon_\sigma(\lambda,\phi,\psi)$ | is the spectral emissivity of the water surface at a position $(\phi,\psi)$ as a function of the nadir angle $\theta$ of the sensor, which depends on the angle $\phi$ and $\psi$, |
| $T_s$ | is the water surface temperature at a position $(\phi,\psi)$, |
| $T_a$ | is the atmospheric temperature, |
| $T_b$ | is the equivalent background-radiation temperature, |
| $\tau_a(\lambda,\phi,\psi)$ | is the spectral transmittance of the atmosphere at a position $(\phi,\psi)$ as a function of the nadir angle, |
| $\lambda$ | is the wavelength, |

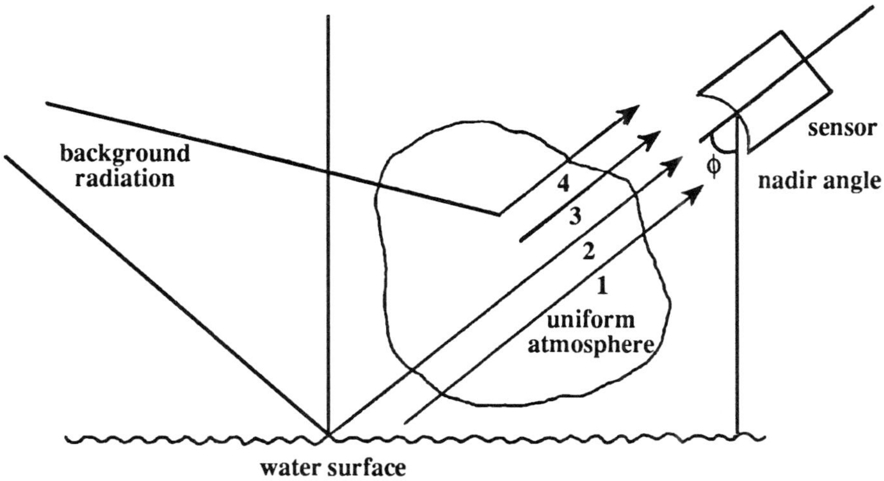


1. Thermal radiation of the water surface.
2. Background radiation reflected at the water surface.
3. Thermal radiation of the atmosphere along the optical path of the sensor.
4. Background radiation scattered in the atmosphere.

Figure 16.1. Passive infrared remote sensing model.

$\phi$ is the horizontal scanning angle on the display of the sensor,

$\psi$ is the vertical scanning angle on the same display,

$L(\lambda,T) = (C_2 / \lambda^5) / [\exp(c_1 / \lambda T) - 1]$ is the spectral radiance of the black body given by the Planck's radiation law,

$c_1$ is the first radiation constant,

$c_2$ is the second radiation constant,

The detector with an optical filter has a responsivity in a spectral interval from $\lambda_1$ to $\lambda_2$ in which $\tau_0(\lambda)$, $\varepsilon_\sigma(\lambda,\phi,\psi)$ and $\tau_a(\lambda,\phi,\psi)$ are independent of $\lambda$. Therefore, the received radiant flux $\Phi_0(\phi,\psi)$ is given by:

$$\Phi_0(\phi,\psi) = (\pi/4)\left(A_d D^2/f^2\right)\tau_0\left\{\left[\varepsilon_\sigma(\phi,\psi)\int_{\lambda_1}^{\lambda_2}\left(L(\lambda,T_s\right)d\lambda\right.\right.$$

$$+ (1 - \varepsilon_\sigma(\phi,\psi))\int_{\lambda_1}^{\lambda_2} L(\lambda,T_b)\,d\lambda\Big]\,\tau_a(\phi,\psi)$$

$$+ (1 - \tau_a(\phi,\psi))\int_{\lambda_1}^{\lambda_2} L(\lambda,T_a)\,d\lambda\Big\} \tag{16.2}$$

We now define

$$\phi_t = (\pi/4)\,(A_d D^2/f^2)\,\tau_o \int_{\lambda_1}^{\lambda_2} L(\lambda, T_i)\, d\lambda \qquad (16.3)$$

where the subscript i becomes o, s, a or b corresponding to observed surface, atmosphere or background, respectively.

Assuming that the observed temperature $T_o$ can be calibrated from the value displayed on the thermal infrared mapper, as discussed in the next section, and that $T_a$ is equal to $T_b$, we obtain the value of $\Phi_s$ $(\phi, \psi)$ to be detected, that is, the radiant flux of thermal radiation from the water surface from equation (2), as follows;

$$\Phi s\,(\phi,\psi) = [\Phi_o\,(\phi,\psi) - (1 - \varepsilon_\sigma\,(\phi,\psi)\,\tau_a\,(\phi,\psi))\,\Phi_b\,(\phi,\psi)\,]\,/\,[\varepsilon_\sigma\,(\phi,\psi)\,t_a\,(\phi,\psi)] \qquad (16.4)$$

The surface temperature $T_s$ at a position $(\phi,\psi)$ can thus be obtained by taking the inverse of equation (16.3).

## *Basic Experiments*

In order to estimate the true temperature of the water surface by infrared remote sensing, three steps are necessary : firstly, calibration of the thermal infrared mapper, secondly, the calculation of the emissivity of the water surface as a function of the nadir angle $\theta$ of the sensor, which depends on the angles $\phi$ and $\psi$, and thirdly, compensation for the atmospheric absorption along the optical path of the sensor. In this study, the sensor, a thermal infrared mapping system, had a HgCdTe element detector and two mechanical scanning mirrors, with specifications shown in Table 16.1.

TABLE 16.1. Specifications of the thermal infrared mapper sensor.

| | |
|---|---|
| Temperature range. | -20 to 1600 °C |
| Minimum detectable temperature difference | 0.2 °C |
| Flame time. | 3.7 sec |
| Field of view | 25 ° in horizontal |
| | 25 ° in vertical |
| Scanning lines | 224 lines |
| Detector | HgCdTe (77 °K) |
| Spectral range | 9.3 to 11.35 μm |
| Range of focus | 20 cm to ∞ |

FUJITSU Co.Ltd. INFRA-EYE 560

The results of the calibration procedure are described below :

(1) The calibration curve of the sensor is shown in Figure 16.2. The sensor temperatures are plotted versus the temperatures of a blackbody-type source located 70 cm in front of the sensor. The blackbody-type source was a cylindrical cavity of 33 mm diameter aperture and 25 mm in depth constructed in the side of an acrylic plastic tank and then the temperature of the blackbody-type source was equivalent to the temperature of the water in the tank. The probability error of the blackbody-type source compared with both temperatures was within 0.18% (Itakura *et al.*, 1989).

(2) The emissivity of the water surface changes with the nadir angle of the sensor. The value of emissivity as a function of nadir angle can be calculated from the refraction index of water, n=1.2, at the infrared wavelength, 10 μm, by the Fresnel theory of reflection at a specular surface. Under conditions in which the nadir angle is less than 60°, the emissivity of water is greater than 0.97; whereas at nadir angle greater than 60°, the emissivity decreases rapidly. At angles of 84° or over, the background radiation overwhelms the sensor. In the infrared wavelength region, the effective optical depth of water is very small, less than 10 μm, due to the strong absorption of radiation by water molecules (Kelly, 1978).

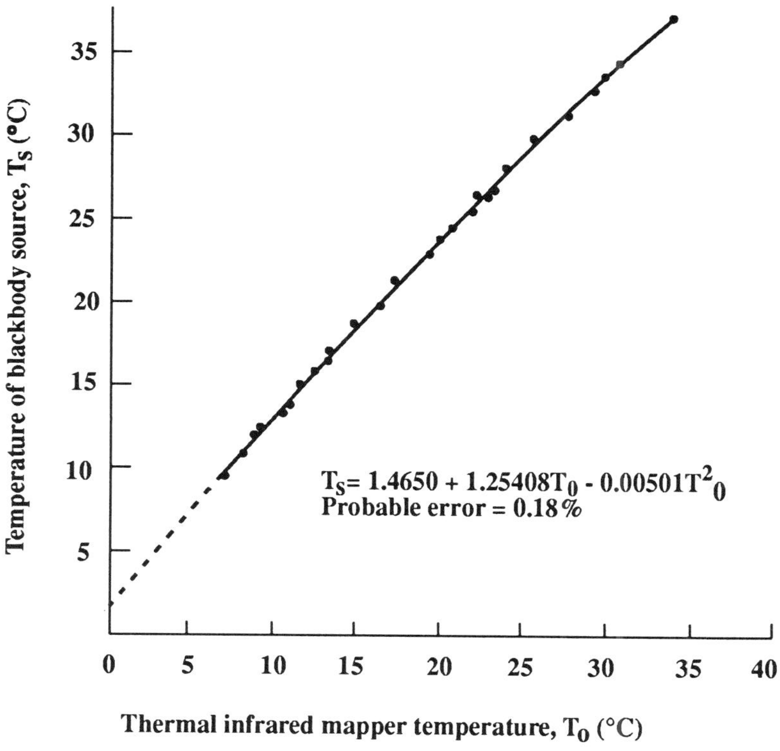


Figure 16.2. Calibration curve of the thermal infrared mapper sensor.

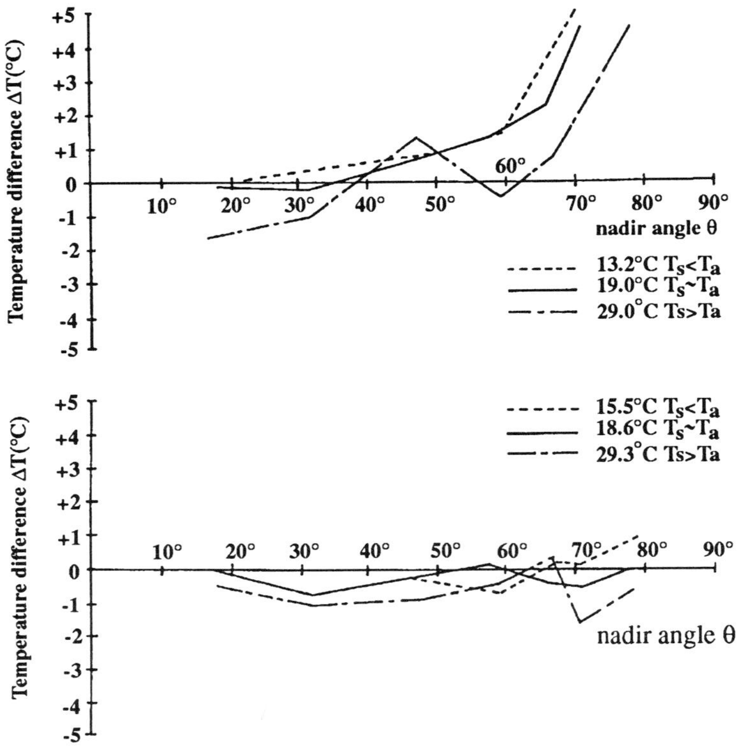


Figure 16.3. Experimental results of infrared radiation thermometry. The difference between the temperature of the blackbody type source and the thermal infrared mapper is a function of the nadir angle.

(3) The atmospheric transmittance along the optical path of the sensor can be estimated by the absorption coefficients of water and carbon dioxide under the meteorological conditions of the observation. Therefore, when the sensor is mounted at the top of a mountain or a high building, assuming a uniform atmospheric condition, it is possible to compensate mathematically for the radiation received due to atmospheric absorption along the slant path. The water temperature was measured in a large acrylic tank with a 1 m diameter opening with the sensor specified in Table 16.1. The nadir angles varied from 18° to 80° corresponding to an optical path between 10 m and 70 m. Figure 16.3 shows the results and they indicate the accuracy of the temperature measurements was within 1°C at nadir angles smaller than 60° during the daytime, and within 0.6°C at nadir angles smaller than 70° at night (Itakura *et al.*, 1989).

Normally, the minimum detectable temperature resolution of the sensor is about 0.2°C, and thus the comparison of temperature between the two relatively large observational areas is of this order. In many cases, these temperature differences can provide useful information about movements of lake water.

# Applications

## *Cyclonic Gyre Generated in the Model of Lake Biwa*

During periods of thermal stratification, Lake Biwa, like most large lakes in the Northern hemisphere, develops a stable gyre, a counter-clockwise surface circulation pattern or a cyclonic gyre (Emery and Csanady, 1973; Csanady, 1977). The surface gyre in North Basin of Lake Biwa has been observed (Suda *et al.*,1926; Okamoto and Motikawa, 1961; Endoh *et al.* 1981) and various theories have been proposed to explain its formation and stability (Endoh 1986) (see Chapter 1). Oonishi (1975) and Okubo *et al.* (1984) theorized that the gyre may be a thermally induced current and tested this hypothesis in a laboratory model of Lake Biwa. The horizontal and the vertical scales of this distorted model, determined on the basis of the Rossby and Ekmann number scaling, are 1/30,000 and 1/300, respectively (Itakura *et al.*, 1990). The model turns at a speed of 1 revolution per minute equivalent to one day for Lake Biwa. Heaters placed along the east and west edges of the North Basin of the model simulate the warming of the shallow coastal regions of Lake Biwa, and thermal infrared mapping was used to quantify the distribution of surface water temperature in the model. The infrared sensor was set 1m above the model of Lake Biwa and the temperatures shown in Figure 16.4 were calibrated with the curve shown in Figure 16.2.

Figure 16.4 shows the changing temperature patterns of the developing gyre as a function of time. Initially, the water surface had a uniform temperature shown in Figure 16.4(a). However, as the heating process began, the water warmed along the fringe areas as shown in Figure 16.4(b). The water surface temperature pattern evolved and the unstable pattern illustrated in Figure 16.4(c) appeared. The flow continued to develop until eventually a fully developed gyre was produced (Figure 16.4(f)). The center contained the lowest temperature, about 2.6°C lower than in the fringe areas and the relative diameter and center position of the gyre in the model are similar to those recently observed in Lake Biwa (see Chapter 3).

## *Water movements in Shiozu Bay*

The circulation in Shiozu Bay discussed in Chapter 4 has been studied for many years in order to develop a physical model describing the water mass exchange between the North Basin of Lake Biwa and Shiozu Bay. The length from north to south is about 1 km, that from east to west is almost 4km and the average depth is 30 m, as shown in Chapter 4. An important aspect of this determination of the circulation is the measurement of surface water temperature. Figure 16.5(a) shows the transect lines for a thermistor towed by a boat in the water and a thermal infrared mapper positioned on the top of a mountain about 220 m above the water surface level of Shiozu Bay. 16.5(b), (c) and (d) show the comparisons of temperatures between the two sensors.

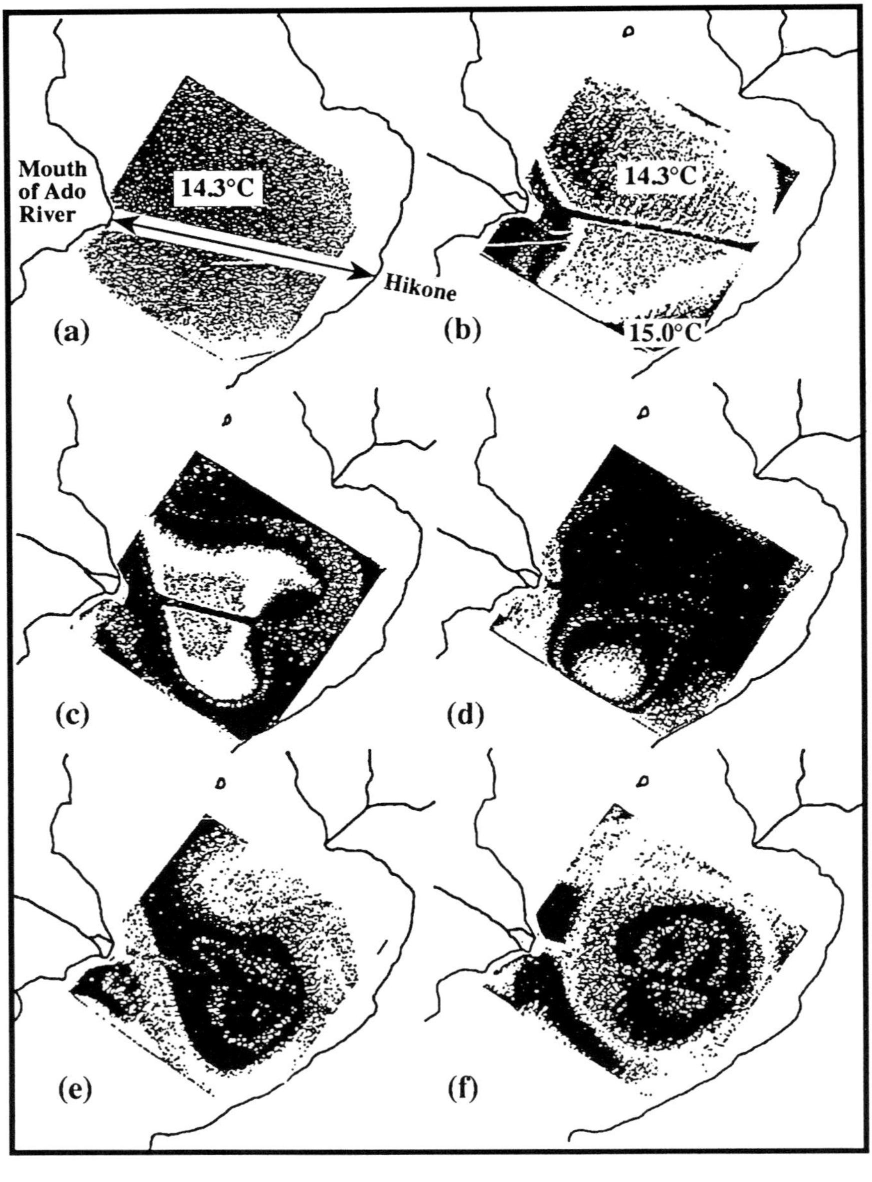


Figure 16.4. Thermal infrared imagery of the gyre generated in the laboratory model of Lake Biwa. The horizontal and the vertical scales of this model are 1/30,000 and 1/300, respectively. (a) the thermal pattern prior to the initiation of the heating process, (b) warming pattern near the fringe area developed after 20 min of heating, (c) unstable pattern developed after 40 min of heating, (d) initial development of the circular pattern after 1 h of heating, (e) the development of the more circular pattern after 1 h 20 min of heating and (f) the development of the fully grown gyre after 1 h 35 min of heating.

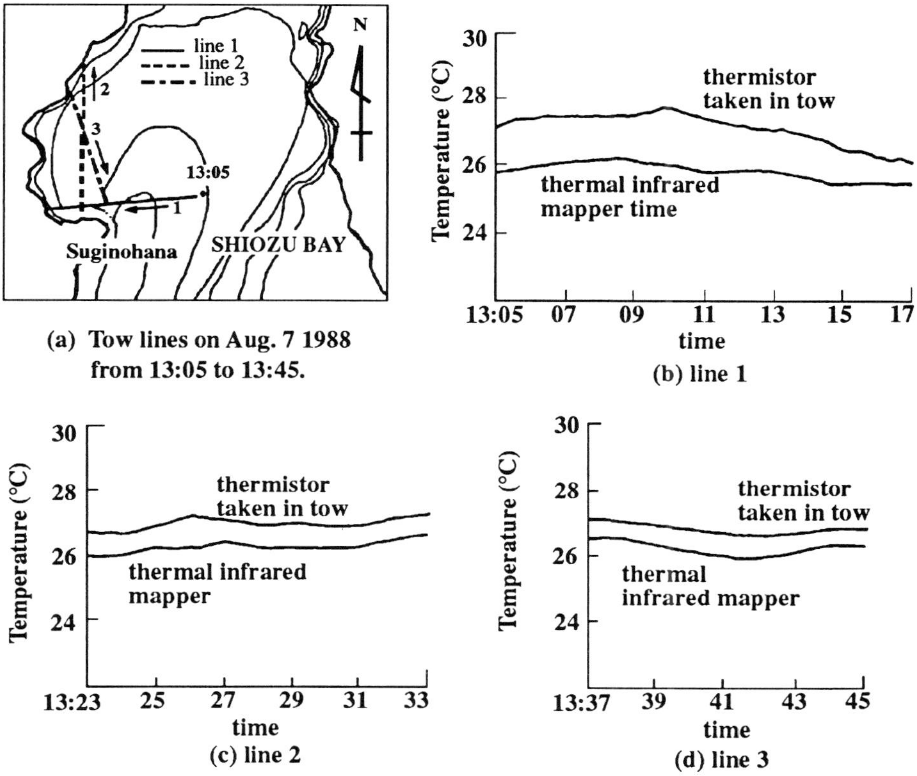


Figure 16.5. Comparison of the surface temperature in Shiozu Bay obtained by two sensors; the thermal infrared mapper and the submerged thermistor.

The temperature data obtained with the thermal infrared mapper sensor were estimated by the calibration method described above in the basic experiment and equation (16.3). The temperature difference was about 1°C, and while the temperature tended to vary in both, the infrared measurement was consistently less than the direct measurement.

A satellite's sensor is reproducable to 0.8°C when the temperature of the sea surface is being measured (Takayama and Maeda, 1990). In the remote sensing study of Shiozu Bay, the accuracy of thermometry was nearly equal to that of a satellite's sensor system, despite the slanted path of the sensor: the shaded area in Figure 16.6(a),(b) and (c) presents the lowest temperature range, 25.0°C to 25.2°C. Figures 16.6(a), (b) and (c) show the variations of the surface water temperature patterns. The patterns suggest the water was flowing toward the north as shown by the thick solid lines in Figure 6(d). This predicted water movement agrees with the actual water movement determined experimentally using a floating flag method and shown by the thin solid lines.

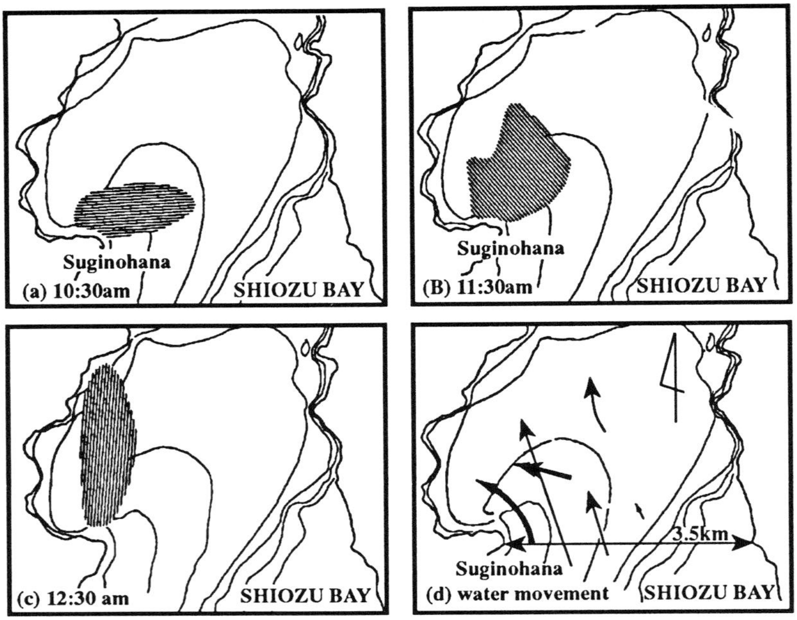


Figure 16.6. Variation of the surface water temperature patterns in Shiozu Bay from 10:30 am to 12:30 am on August 9, 1986 and the water movement patterns inferred from them. The shaded area is the lowest temperature range, 25.0°C to 25.2°C.

## *Water Movement at the Mouth of the Seri River*

A front may form at the boundary between two water masses such as near the confluence of a river with a lake. The mixing of river water flowing into a lake begins as soon as the river water makes contact with the lake water. In addition to the physical mixing phenomena, biological and chemical reactions can also occur in these regions, such as phytoplankton blooms. These phenomena affect the water quality of the lake, therefore methods for observing the distribution and variation of the surface water temperature frontal behavior and mixing processes around the river mouth are of great importance.

Studies of this type have been conducted since 1985 near the mouth of the Seri River. Figure 16.7 shows the variations in surface temperature patterns on October 21, 1987 between 11:34am and 11:44am. A low temperature water mass area (9.5-10.0°C) appeared in (a) and two such areas were in (b). The low temperature areas moved from the offshore area into the mouth of the Seri River, as shown in (c) and (d), and disappeared in (e) and (f). Figure 16.8 shows the water current vectors obtained from the infrared mapping data. The solid lines show the current vectors from 11:15am to 11:17am; the speeds were 22 cm $s^{-1}$ at point A1 and 27 cm $s^{-1}$ at point A2, respectively. On the other hand, the dotted lines present the vectors from 11:41 to 11:44; the speeds were 25 cm $s^{-1}$ at A1 and 42 cm $s^{-1}$ at A2, respectively. At the same time, experimental

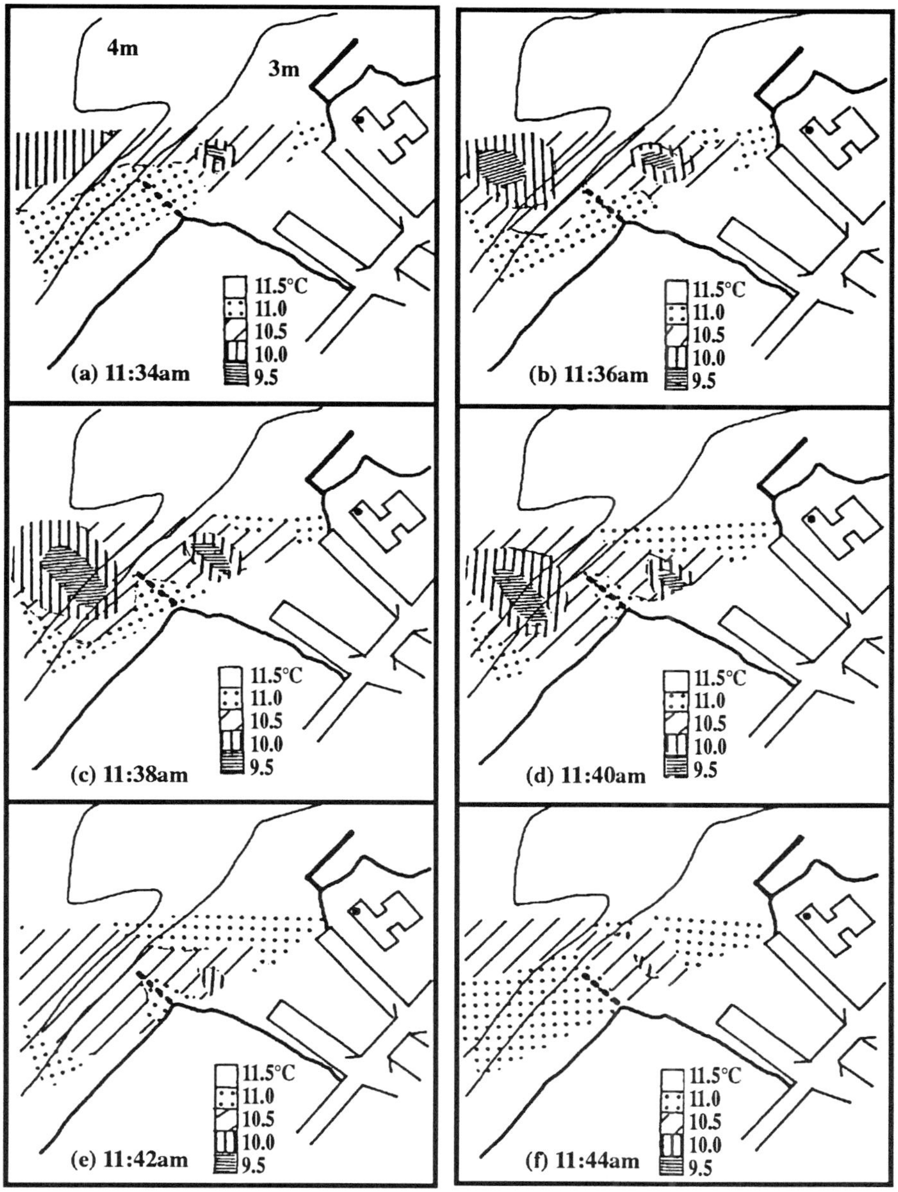


Figure 16.7. Variation of the surface water temperature patterns near the mouth of the Seri River from 11:34 am to 11:44 am on October 21, 1987.

observation with current meters indicated that the current was caused by a strong northwest wind with a velocity of about 10 cm $s^{-1}$.

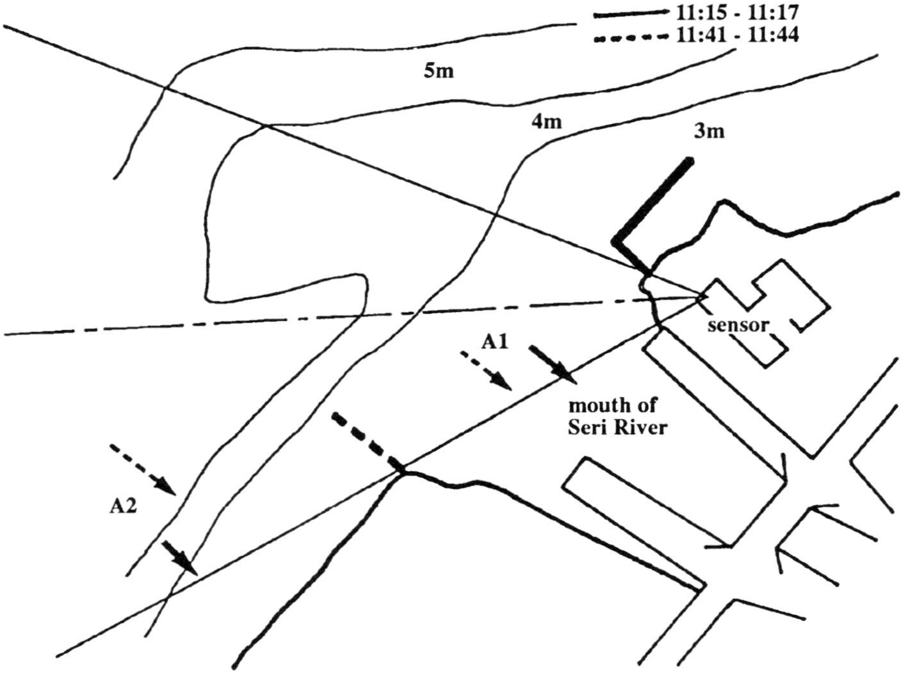


Figure 16.8. Mean current speed and direction (estimated by the thermal infrared imagery) near the mouth of the Seri River from 11:15 am to 11:17 am (solid line) and from 11:41 am to 11:44 am (dotted line) on October 21, 1987.

# Conclusion

Infrared remote sensing is a useful technique for real-time determination of water movement in lakes based on the variation of the distribution of surface water temperature. In the study of Lake Biwa it was established that, in spite of the slanted observation path of the sensor, the accuracy of the temperature field (about 1°C) determined by the thermal infrared mapper sensor is nearly as high as that of the satellite's sensor looking directly down upon the water surface. While this accuracy is not sufficient to develop a numerical model to describe the dynamics of lake water and atmosphere interaction, the variation of the surface temperature pattern can be used to estimate the movement of the surface. The predicted movement provides useful information about chemical, biological and physical parameters of importance to the productivity of an aquatic ecosystem. Thus, infrared remote sensing can be expected to contribute to the development of environmental science and engineering, and to the mapping of water currents within aquatic ecosystems.

# References


Csanady, G. T., (1977) On the cyclonic mean circulation of large lakes. *Proc. Nat. Acad. USA*, 74(6), 2204-2208.

Emery, K. O. and G. T. Csanady, (1973) Surface circulation of lakes and nearly land-locked Seas., *Proc. Nat. Acad. USA*, 70(1), 93-97.

Endoh, S., I. Okamoto. and M, Nakai, (1981) Circular currents in the North Basin of Lake Biwa, (1)., *Jap. J. Limnol.* 42, 144-153 (in Japanese).

Endoh, S., (1986) Diagnostic study on the vertical circulation and the maintenance mechanisms of the cyclonic gyre in Lake Biwa. J. *Geophy. Res*. 91(C1), 869-876.

Itakura, Y., T. Kawasaki, M. Hachida, S. Morikawa, H. Miyagawa, S. Endoh, S. Kitamura, and I. Okamoto, (1989) Infrared remote sensing for water currents of Lake Biwa, (1). *Memoirs of the Faculty of Education, Shiga Univ*. 39, 77-85 (in Japanese).

Itakura, Y., S. Endoh, I. Okamoto, and M Kumagai, (1990) Report of the 54th Conference of the Japanese Society of Limnology II Lake Biwa, *Jap. J. Limnol*. 51, 26-28.

Kelly, C. S., (1978) Effective infrared optical depth asssociated with the clear ocean. *Appl. Opt,* 17, 3054-3059.

McClain, E. P., (1985) Comparative performance of AVHRR-based multichannel sea surface temperatures. *J. Geophy. Res*. 90(C6), 11587.

Okamoto, I. and M. Motikawa, (1961) Water circulation in Lake Biwa as deduced from the distribution of water density. *Jap. J. Limnol*. 22, 193-200 (in Japanese).

Okubo, K., Y. Muramoto, Y. Oonishi, and M. Kumagai, (1984) Laboratory experiments on thermally induced currents in Lake Biwa. *Bull. Disas. Prev. Res. Inst.,* Kyoto Univ., 34(2 No.304), 19-53

Oonishi, Y., (1975) Development of the current induced by the topographic heat accumulation, (1). *J. Oceanogr. Soc. Japan*, 31, 243-354.

Takayama, Y. and K. Maeda, (1990) Sea surface temperature observation by VTIR on MOS-1 satellite. *Proc.Infrared Soc. of Japan* 17, 35-40 (in Japanese).

Tameishi, H., (1985) Sea surface temperature observation and fish-finding by satellite sensor. *Proc.Infrared Soc. of Japan* 10, 70-83 (in Japanese).

Suda, K., K. Seki, J. Ishii, S. Takaishi, and M. Mizuuchi, (1926) The report of limnological observation in Lake Biwa, (1). *Bull. Kobe Marine Observatory* (in Japanese).

# 17

# Laser Sensing of Phytoplankton Size

R. Tsuda, Y. Kakui and M. Kumagai

## Introduction

The effects of aquatic eutrophication in coastal and inland waters, such as lakes, is an increasing problem. Outbreaks of red tide, for example, lead not only to a deterioration in water quality, but also threaten human life. Red tides occur at a time of extremely rapid propagation and accumulation of a specific phytoplankton, and it is therefore important to be able to predict vertical and horizontal variations in concentration (cells or colonies) of species. In general, techniques are needed to obtain not only knowledge related to the size structure of phytoplankton, but also to gain information on the diversity of species. Techniques that have been used include an optico-electronic system (Cooke *et al.*, 1970), particle counters based on electrical principles (Sheldon *et al.*, 1972; Pugh, 1977) and instruments based on optical principles using laser technique (Bale and Morris, 1987; Spinrad and Yentsche, 1987; Cunningham, 1989). For in situ measurements of phytoplanktonic size spectra, Kakui *et al.* (1985) have developed a new instrument employing optical fiber cables and a laser technique to measure the concentration of phytoplanktonic particles in equal size ranges. Practical application demanded certain improvements, the success of which has been reported before (Tsuda *et al.*, 1989; Nakata and Ishikawa, 1989; Tsuda *et al.*, 1990). The present study has been carried out using this new technique for characterizing size spectra of phytoplankton from in situ measurements in Lake Biwa.

Physical Processes in a Large Lake: Lake Biwa, Japan
Coastal and Estuarine Studies, Volume 48, Pages 195–201

# Method

## *Principles of the Measuring System*

The measuring system is composed of two subsystems (Figure 17.1). The shipboard unit consists of a laser light source and data processors, while the underwater unit contains the detecting optics. The two subsystems are connected by an optical fiber cable. Because it contains only optics, the compact submersible unit is ideally suited to towing. A laser light source (Ar-ion laser of 488 nm wavelength) is focused into a stream of water. When the phytoplanktonic particles cross the irradiated spot in the water, chlorophyll-*a* fluorescence occurs and the peak height of these emitted optical pulses may be correlated with the cross-section of phytoplanktonic particles (Tsuda and Kakui, unpublished). The pulse-height distribution is transformed into a size distribution and the concentrations of phytoplanktonic particles in ranges of equal size are obtained from the counts (Figure 17.2), using the method of data processing as described by Kakui *et al.* (1985).

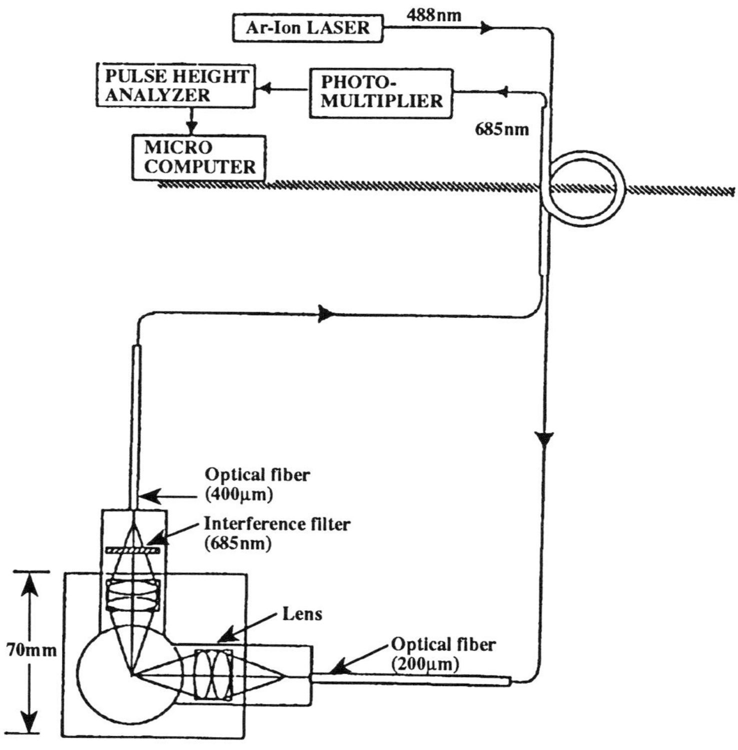


Figure 17.1. Block diagram of measuring system.

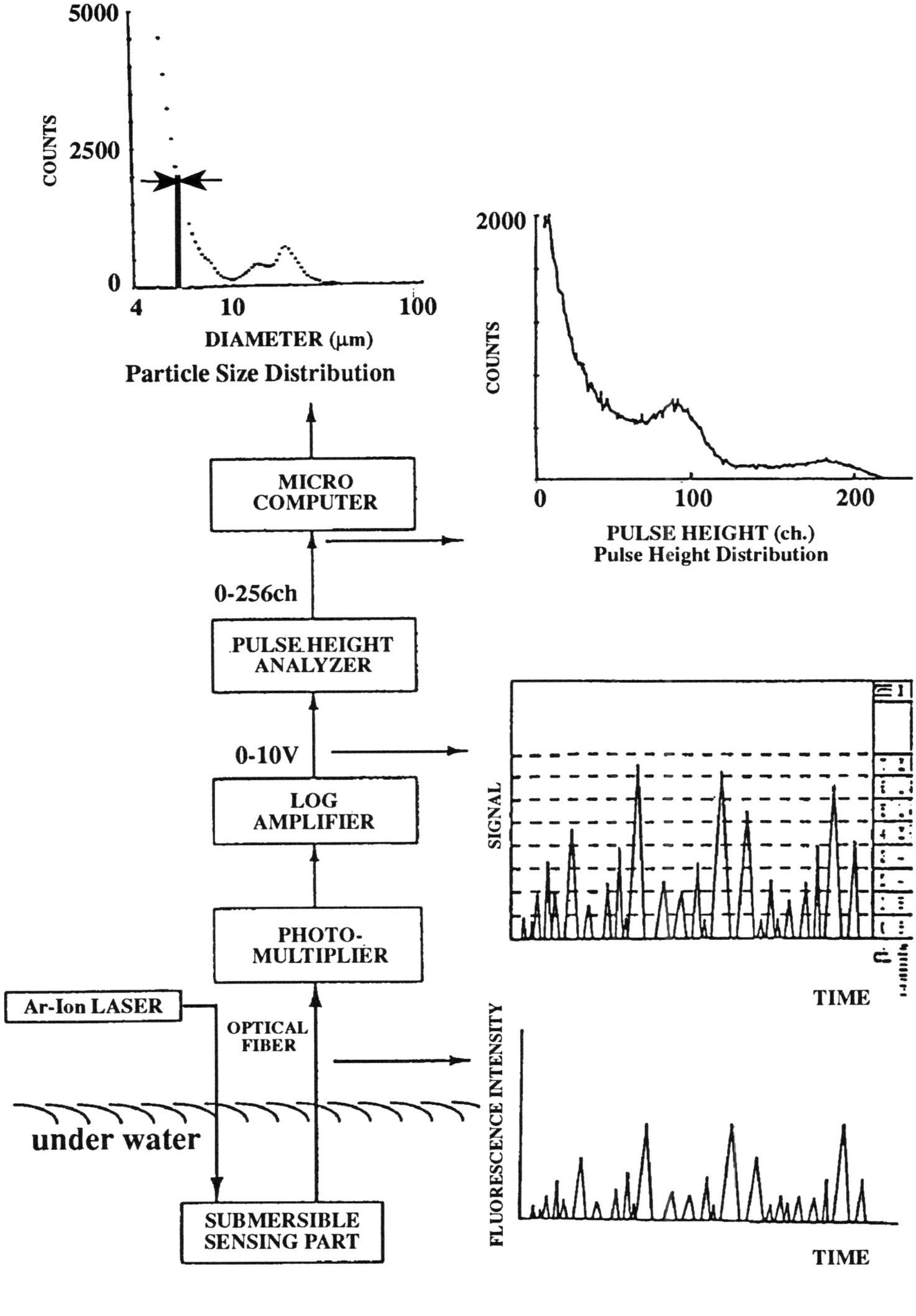


Figure 17.2. Scheme for analyzing size spectra of phytoplanktonic particles.

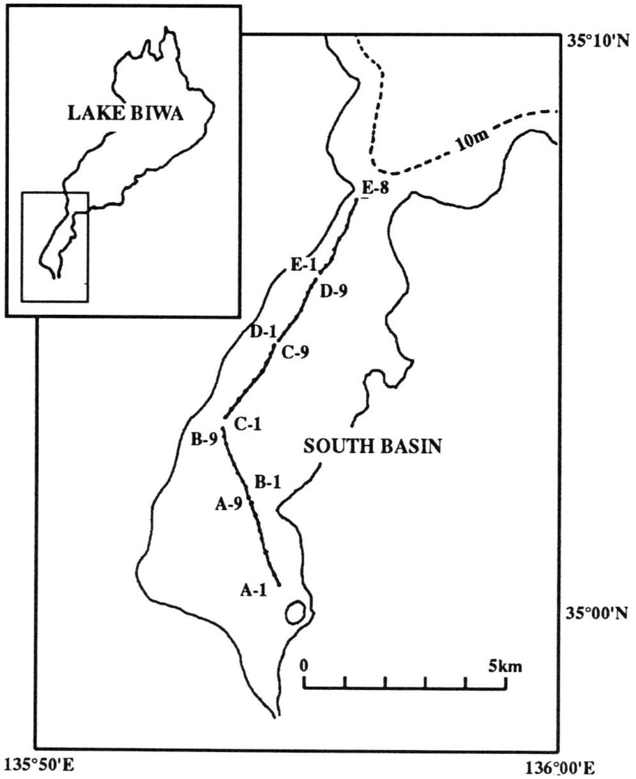


Figure 17.3. Station locations in the South Basin of Lake Biwa, used on September 22, 1988.

# Results

## *Phytoplanktonic Size Spectra*

Figure 17.3 shows a transect between St. A-1 and St. E-8 in the South Basin. The apparatus was fixed at a depth of 0.5 m depth and towed at a speed of almost 2.0 $ms^{-1}$. Counts were made over 100 second periods throughout the transect and simultaneous measurements of water temperature and light attenuation were also recorded. Figure 17.4 shows examples of phytoplanktonic size spectra. The data from St. A-8 indicates that there are two peaks at around 5 μm and 25 μm equivalent diameter. According to microscopic observations, the phytoplankton community for the sampled waters was dominated by the diatom *Melosira granulata* and the size of this species lay between 12.0 and 30.2 μm equivalent diameter (averaged diameter of 20.2 μm), accounting for the counts of larger size. However, dominant species of smaller size were not found by

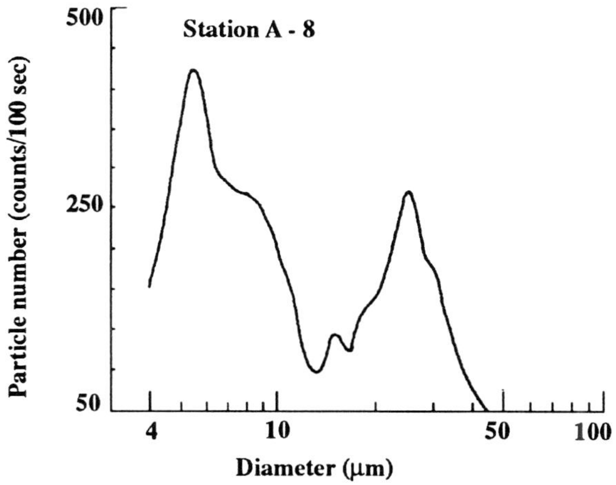


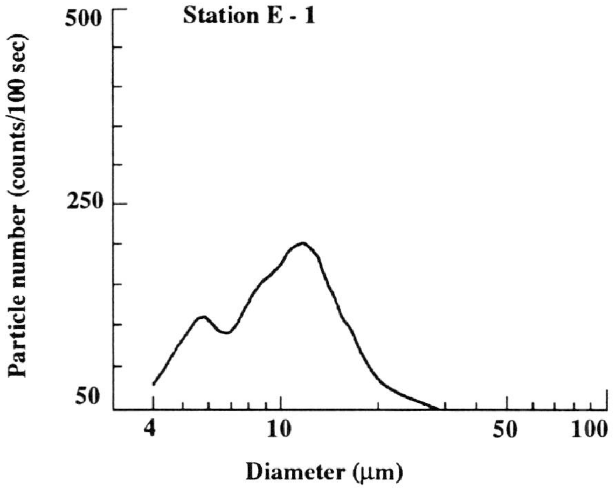


Figure 17.4. Size spectra of phytoplankton computed from the system of particle-size analysis.

microscopic observations. The smaller peak at 5 μm also occurs at St. E-1 and it appears that counts of this size were due to the number of nanoplankton and possibly cells of disjointed colonies. The dominant species at St. E-1 were the diatom *Staurastrum dorsidentiferum* and the blue-green alga *Gomphosphaeria lacustris* with 15 μm and 35 μm equivalent diameter, respectively. The peak at around 15 μm of size spectra results from counts dominated by *S. dorsidentiferum*, but there was no evidence for a peak corresponding to the presence of *G. lacustris*. This species is apt to be disjointed with the result that the sensor may be measuring the fluorescence radiation which is scattered in individual cells.

## *Horizontal Distribution of Size Spectra*

Figure 17.5 shows the records of water temperature, light attenuation and counts of phytoplanktonic particles along Lines A-1 to B-9. Large variations occurred in counts of phytoplanktonic particles near St. A-9 and St. B-1 and, in general, there are corresponding variations in the light attenuation. A front appeared to have been crossed as there was also a small temperature decrease and the rapid decline in particle counts could be associated with known changes in the phytoplankton population. However, for this particular transect, it is clear that there is much small-scale spatial variability in the specific phytoplankton population. We can obtain variations in chlorophyl *a* fluorescence from the continuous measurements, but examination of Figure 17.5 reveals that it is impossible to extract information on the size spectra of phytoplankton because this data shows only the concentration of phytoplankton. Figure 17.6 shows the spatial changes of size spectra throughout the transect in the South Basin. The spatial distribution of size spectra varied drastically between Line A-9 and Line B-1. The size distribution along Line A is characterized by two peaks of around 5 μm and 25 μm equivalent diameter, and the dominant species *M. granurata* is the main contributor to a peak of larger size mentioned above. We were unable to find the dominant species of smaller size, but the two peaks appear to be associated with different populations. Size distribution along Line B-1 to B-9 gradually changed in shape. Decreased counts of the smaller and larger size saw the two peaks of around 5 μm and 20 μm vanish and the size spectra at Line B-9 displayed only a peak of around 10 μm. It appears that there is significant variation with species diversity.

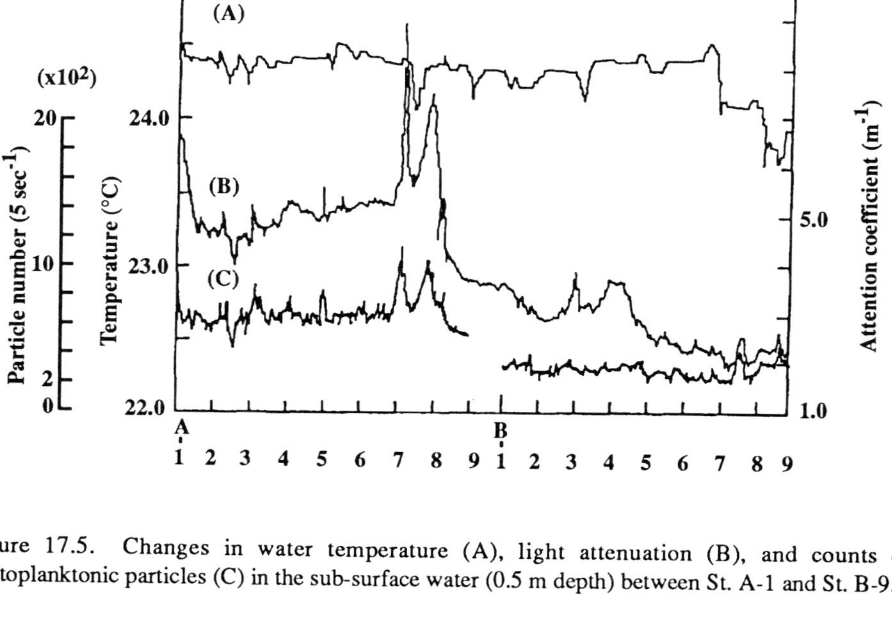


Figure 17.5. Changes in water temperature (A), light attenuation (B), and counts of phytoplanktonic particles (C) in the sub-surface water (0.5 m depth) between St. A-1 and St. B-9.

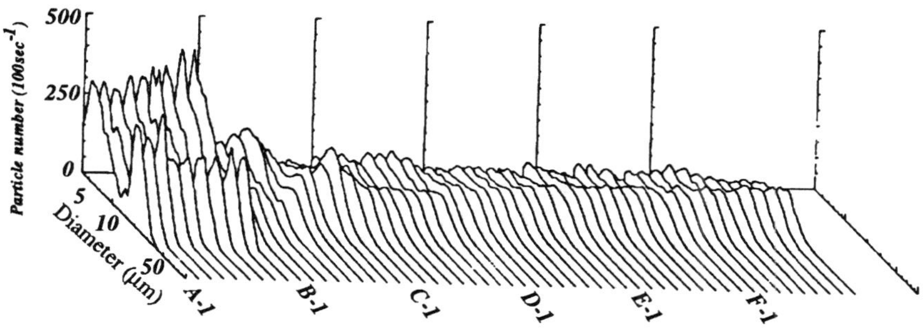


Figure 17.6. Horizontal distribution of size spectra throughout (St. A-1 to St. E-8).

# Conclusion

From this new instrument, useful data related to the size and concentration of phytoplankton can be derived from a wide area in a short time. It should therefore be possible to estimate the biomass of phytoplankton in Lake Biwa. In addition, the new instrument should be able to predict the propagation of *Uroglena americana*, the 'red tide', which frequently occurs in Lake Biwa in May or early June and may accelerate the eutrophication process.

# References


Bale. A. J. and A. W. Morris, (1987) In situ measurement of particle size in estuarine waters. *Estuar. Coast. Shelf Sci.*, 24, 253-263.

Cooke, R. A., L. D. B. Terhune, J. S. Ford and W. H. Bell, (1970) An optico-electronic plankton sizer. *Fish. Res. Bd. Can. Tech. Rep*. 172, 40.

Cunningham, A., (1989) A low-cost, portable flow cytometer specifically designed for phytoplankton analysis. *J. Plankton Res.*, 12, 149-160.

Kakui, Y., A. Nishimoto, J. Hirono, and M. Nanjo, (1985) Underwater analysis of suspended biological particles with an optical fiber cable. *Adv. Chem. Seri.*, 209, 275-292.

Nakata, K. and K. Ishikawa. (1989) In situ measurement of phytoplankton size spectra in a coastal region using optical fiber sensor. *Pro. Adv. Mar. Tech. Conf.,* 1, 65-70.

Pugh, P. R., (1977) The application of particle counting to an understanding of the small-scale distribution of plankton. In: *Spatial Pattern in Plankton Communities*. edited by J. H. Steele,.111-129. NATO Conf. Seri., Series IV: Marine Scinces.

Sheldon, R. W., A. Prakash and W. H. Sutcliffe, (1972) The size distribution of particles in the ocean. *Limnol. Oceanogr*., 17, 327-340.

Spinrad, R. W. and C. M. Yentsch, (1987) Observations on the intra- and interspecific single cell optical variability of marine phytoplankton. *Appl. Opt.* 26, 357-362.

Tsuda, R., Y. Kakui and T. Saito, (1989) Observations of marine phytoplankton size using optical fiber cables with the laser technique. *Pro. Adv. Mar. Tech. Conf.*, 1, 59-63.

Tsuda, R., Y. Kakui, T. Saito. and M. Kumagai, (1990) In situ measurement of phytoplankton size spectra by laser technique. *Jpn. J. Limnol.*, 51, 28-29.

# 18

# New Measurement Techniques and Some Depositional Problems at the Lake Bottom

K. Yokoyama and S. Okuda

## Introduction

One hundred and ten rivers run into Lake Biwa and large amounts of sediment are transported into Lake Biwa by the rivers each year. The sedimentation rate in the lake bottom has been estimated to be approximately 1 mm per year, although it varies from place to place depending upon the topographic and hydrological features of the water region. The sedimentation rate, particulary near the river mouths, fluctuates with variations in river discharge. While the depositional process has been observed in many lakes, it is usually difficult to make precise estimates of deposition. When considering depositional processes in shallow regions, much more attention needs to be given to the resuspension of lake bottom sediment in order to accurately assess the true fluxes of materials.

Suspended particles precipitate throughout the lake water, however the sediment particles may subsequently be resuspended and then moved again a number of times. The interaction between the lake water and bottom deposits is important, particularly with respect to various environmental problems. For example, resuspension of bottom particles by the wind returns nutrients from the bottom sediments to the lake water. It is necessary to clarify these effects and to correctly estimate the mass flux of various substances from the shore to the central parts of a lake. Thus, in sedimentology and geomorphology much attention has been paid to entrainment processes of bottom deposits and their effect on the sediment structure. In Lake Michigan (Edington and

Physical Processes in a Large Lake: Lake Biwa, Japan
Coastal and Estuarine Studies, Volume 48, Pages 203–213

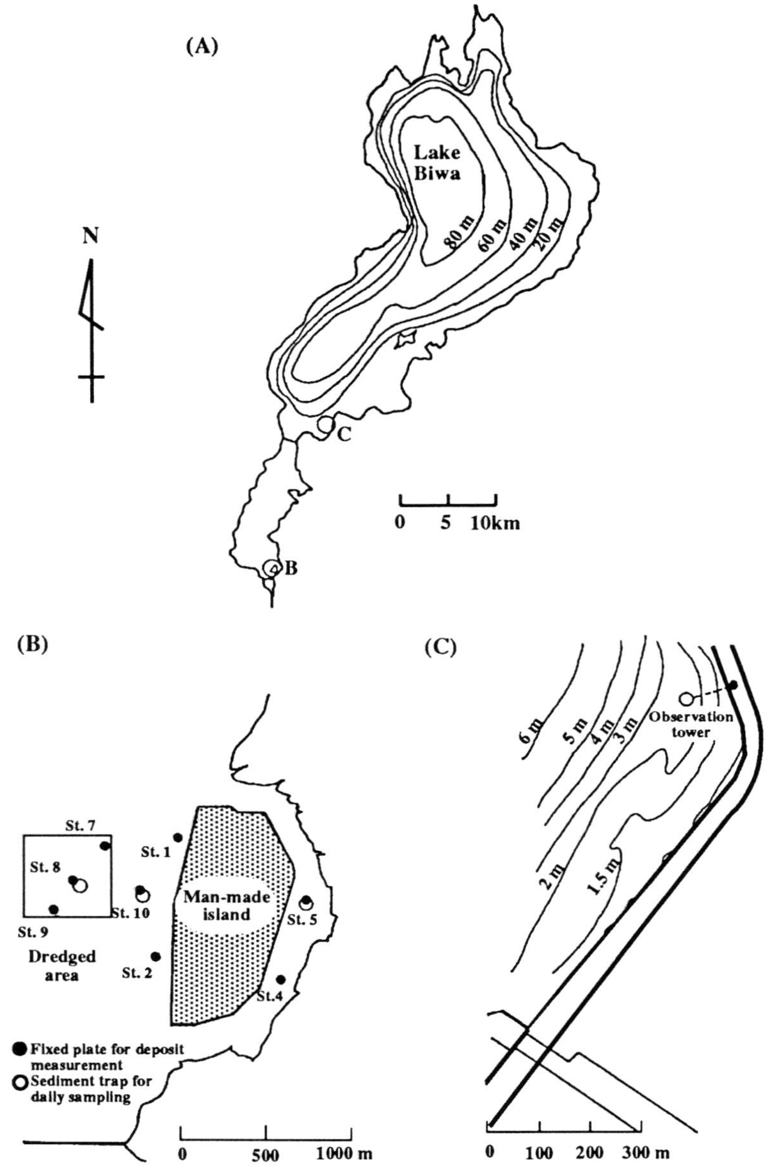


Figure 18.1. Map of the observation area in Lake Biwa.

Robbins, 1976) and Lake Biwa (see Taishi, Chapter 13), storm events can lead to strong erosion of the shoreline or lake bottom, resulting in subsequent deposition of as much as a few centimetres. It has become clear that entrainment and remobilization of bottom deposits plays a prominent role in the depositional processes of a lake. It is necessary for us to establish a proper understanding of depositional processes by means of field observations and measurements.

Measurement techniques concerned with entrainment or removal of bottom sediments in lakes have been developed in the field of coastal engineering, with emphasis on the movement of suspended sediment near the coastline. These techniques, however, can only measure the changes over a finite period, and cannot be used to measure continuous change. With this in mind, we carried out field observations at Yoshikawa and in the southern part of Lake Biwa (Figure 18.1) using a thermistor deposit meter and measuring plate especially developed for this kind of observation.

## Principles of Measurement

### *Thermistor Deposit Meter*

Deposition or erosion rates depend upon the magnitude of wind-generated waves and currents. Deposition rates are usually quite small, except at river mouth areas during floods, and continuous observation is necessary in order to follow the short-term changes in the lake bottom. For this purpose, a thermistor deposit meter was developed and the principles on which this device operates are shown in Figure 18.2. The resistance of a thermistor (R) changes as a function of temperature ($T_a$) and the characteristic relation of R and $T_a$ is shown by the curve in Figure 18.2. The heat produced per unit time by the thermistor is $RI^2$ with a constant electric current, I. The heat flux from the thermistor, at temperature $T_a$, to the ambient environment at temperature $T_b$ is $C(T_a-T_b)$ (where C is the heat transfer coefficient). Since C differs between mud and water, the position can be detected by measuring the resistance of the thermistor. In a steady state, heat flux is equal to heat production (i.e. $R = C\ (T_a - T_b)/I^2$) and this relation is shown as $T_a$-R in Figure 18.2. In this case, the equilibrium point of the thermistor corresponds to point P (shown in Figure 18.2) where the straight line of $T_a$-R crosses the characteristic curve. In this state, the temperature is $T_a$ and resistance is R.

Expressing the position of the thermistor by an index (w) in water and (m) in mud and assuming that the temperatures of the water and mud are almost identical, we can set $T_{bw} = T_{bm} = T_b$. On the other hand, the heat transfer coefficient C satisfies the relation Cw > Cm. Thus, the equilibrium points for w and m are shown in Figure 18.2, with the relative positions $T_{am} > T_{aw}$ and the resistance of the thermistor $R_w > R_m$. The voltage of a thermistor with constant current (1mA and about 0.4 mA in the case of our field measurements in the lake) will satisfy the relation $V_m > V_w$ where $V_m$ and $V_w$ are the voltages of the thermistor in water and mud, respectively. By measuring the voltages at each thermistor level, we can determine whether the particular thermistor is in water or in mud.

The thermistor deposit meter sensors consist of 50 probes fixed on a bar every 5 mm.

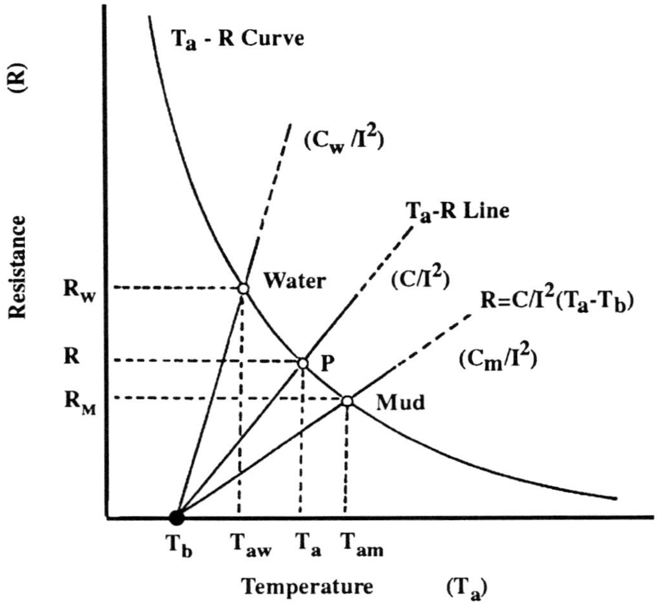


Figure 18.2. Basic principles of thermistor deposit meter. (E) electric current, (C) Heat transfer coefficient, ($T_a$) Thermistor temperature, ($T_b$) Water or mud temperature.

Each thermistor is read in turn by running an electric current to the thermistor for 20 seconds, and then after a lag of 2 seconds, moving on to the next thermistor. If there is a large difference in voltage between two neighboring thermistors, the position of the bottom surface can be determined at a point between the two thermistors. The measurement can be done satisfactorily both under mud and in water, with all thermistors possessing the same characteristics, and bottom-level changes larger than 5 mm can be detected.

## *Plate for Measurement of Deposit Thickness*

A method of measuring deposit thickness by a combination of two plates is shown in Figure 18.3. A plate at a standard level is fixed horizontally in the bottom by a diver (Figure 18.3a). The diver brings a measuring cover (Figure 18.3b) and measures the vertical distance between the fixed plate and the cover. The cover is set along guide bars standing vertically on the surface of the fixed plate. Three holes are used to insert measuring pipes in the cover, with the diver inserting these pipes vertically until they reach the fixed plates and then fixing them to the measuring cover. The cover and fixed

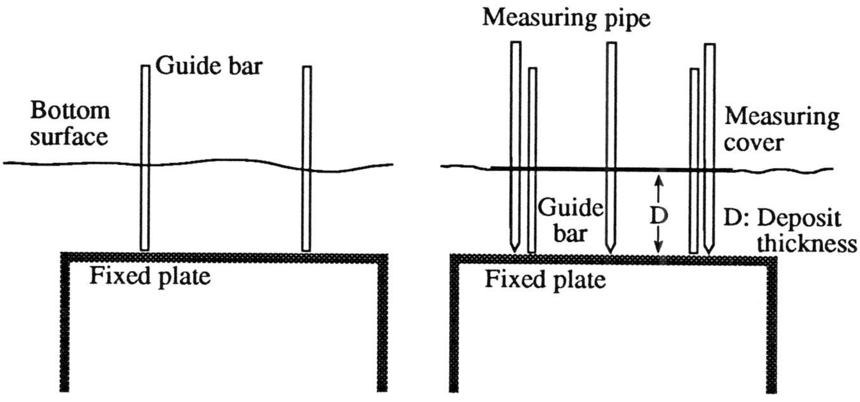


(a) Setting of the fixed plate in the bottom plate

(b) Measurement of the deposit thickness by the measuring cover

Figure 18.3. Plate for measurement of the deposit thickness.

pipes are pulled up to the boat and the lengths of the three pipes (shown as in Figure 18.3b) are averaged to yield the distance D between the fixed plate and cover, equal to the mud layer over the fixed standard plate. Two measurements at a known time interval then yield a measure of deposition. The measuring cover is made with a specific weight equal to water in order to avoid sinking into the mud at the bottom. Test experiments using this apparatus were carried out at several points in Lake Biwa with results suggesting that observational errors lie within ± 2 mm. When these results are compared with those of daily sediment traps, explained hereafter, it is possible to get the net amount or thickness of deposit in a certain period.

## *The Daily Sediment Trap*

This apparatus was designed to measure the flux of suspended sediments, collect samples to examine the quantity of sediment particles, and to check the variation of particles under different hydrological conditions. As shown in Figure 18.4, the apparatus contains 30 bottles set around a circle with a sampling disk which rotates once a day. The hole on the top of this apparatus sends settling particles into the container. In this way, 30-day continuous sampling can be carried out automatically. The amount of sediment obtained by this method shows only accumulation, since sediment once trapped in this apparatus can never be removed.

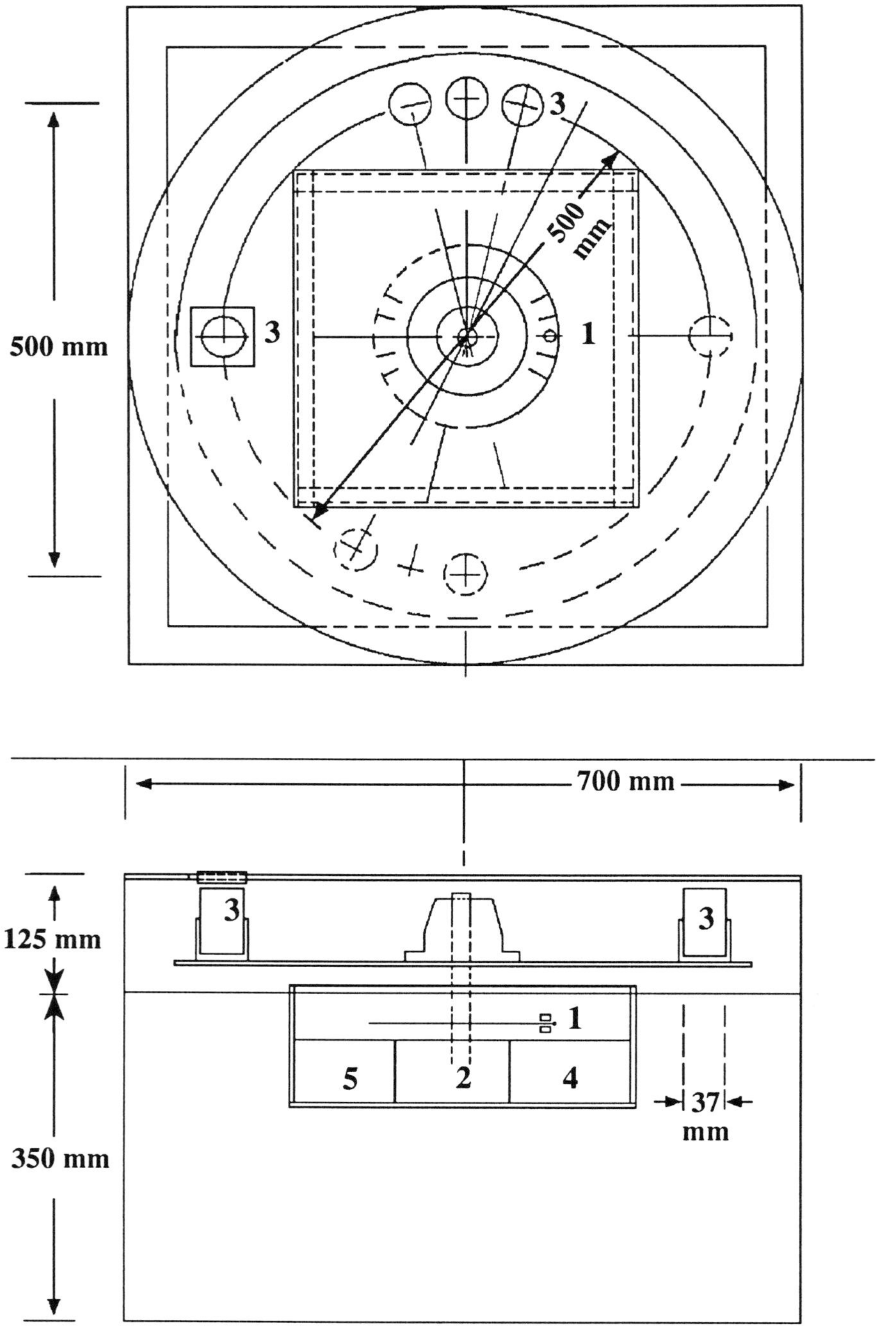


Figure 18.4. Daily sediment trap. (1) photo-electric device (2) driving mechanism, (3) sampling bottle, (4) electric power, (5) I.C. module.

## The Results of Observations

### *Results of Continuous Measurement of the Change in Bottom Surface Levels and Deposit Sampling with the Daily Sediment Trap*

The results of observations using a thermistor deposit meter are shown in Figure 18.5. Since the sensor probes of thermistors are fixed at a 5 mm intervals, the resolving power of this apparatus is over 5 mm. The field site for these measurements is shown in Figure 18.1 (the distance from the lake shore was 100 m and the water depth 1.8 m). The main sensor was set in this area and the other sensors set at the lake shore. The measurements were taken at one hour intervals and at the same time wave height, current velocity, water temperature, electrical conductivity, wind direction and wind speed were also measured at an observation tower set in the area where the sensor was located.

Although these observations had been carried out over a 6 month period that included a high wave period (the maximum wave height being 76 cm ), there was no change in the record of the thermistor deposit meter and so the change in the bottom level was estimated at less than 5 mm for the period. In the same period, glass bead particles painted in different colors were put on the bottom surface to survey the maximum erosion depth. 25 cores near the initial point were sampled at constant intervals. The results showed that the beads remained at a level within the surface layer at a depth not exceeding 3 mm, supporting the observational results of the thermistor deposit meter at the same point.

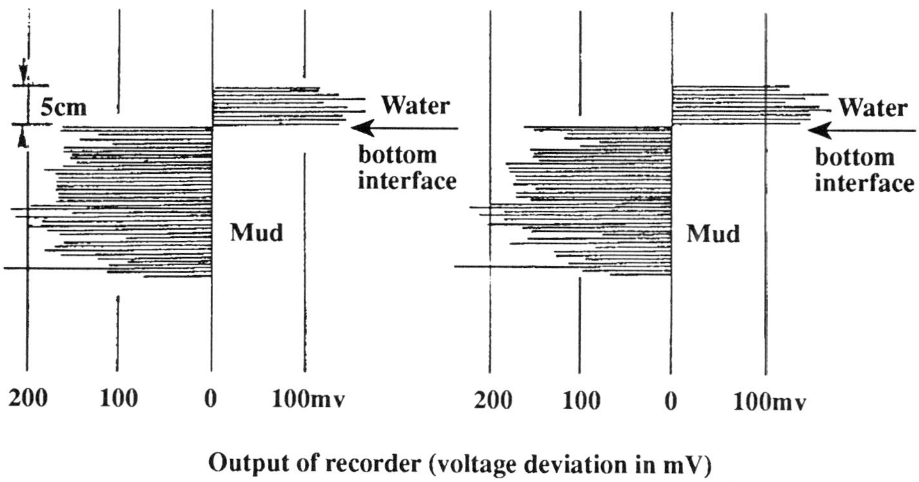


Figure 18.5. Recording measurements on the boundary between the water and the bottom using a thermistor deposit meter.

Using these two different methods, all results proved to be less than 5 mm throughout the observation period.

The daily sediment trap caught the settling particles every day, and the sizes and chemical properties of the sediments differed in several ways. The ratio of coarse to fine sand was great when a large sample amount had been collected, while the organic substance content of muddy materials was larger when only a small sample amount had been collected. This means that large mineral particles settle during flood or high wave periods, while small organic particles are flushed away in that period.

## *The Results of Lake Bottom Change Obtained with a Deposit Thickness Measuring Plate*

Field observations of bottom surface variation have been carried out for several years around a man-made island (shown in Figure 18.1b). Some of these results are shown in Figures 18.6 and 18.7. Deposit thickness measuring plates were set at 10 locations, shown as Stations 1 to 10 in Figure 18.1b. Of these, 3 stations which have different sedimentation characteristics were chosen for the determination of the daily sedimentation characteristics (Station 5, water depth 2 m; Station 8, water depth 2.5 m; Station 10, water depth 14 m; see Figure 18.1b). The sedimentation rates obtained by the deposit thickness measuring plate are shown in Figure 18.6. The values vary from 1 to 7 cm per year and are one order of magnitude higher than those in the middle of the South Basin, away from the island. It seems that depositional processes are more active in winter, and that great variation occurs depending on location, period, and hydrological characteristics.

The daily rate of deposition for every month in summer, winter, and through the observation period are shown in Figure 18.7. The mud accumulation depth (shown in Figure 18.7) is similar to the variation of deposit thickness, being larger in summer than in winter. However, at Station 8, the opposite effect was noted with the mean daily rate of deposition being larger than at other stations. At Station 8, the water depth is 14 m and a density stratified layer appears in the summer that prevents suspended sediment from sinking to deeper layers. At Station 8, the net deposition is larger than at Station 5 in the narrow waterway, because of the water depth and the larger fetch distance of north western seasonal winds.

The seasonal variation of depositional processes are shown in Table 18.1 from the observational results at Station 5, 8 and 10. <T> is the total amount of settling in each period (dried weight) measured by the daily sediment trap, <D> is the deposit thickness measured by the deposit thickness measuring plate, <N> is the dried weight calculated from <D> using the water content and specific weight of the surface mud, <R> is the removal rate calculated by <T> - <N>, and <R>/<T> is removal coefficient, i.e. the ratio of removal rate to total settling rate. The coefficient values show that most settling particles in shallow water regions in our test field have been removed by flushing action during flood and strong wind periods.

TABLE 18.1. The amount of variation of deposit depth and deposition rate in the summer and winter

| Station No. | Station 5 | | Station 8 | | Station 10 | |
|---|---|---|---|---|---|---|
| 1982 | Jan-Mar/July-Sept | | Jan-Mar/July-Sept | | Jan-Mar/July-Sept | |
| Total deposit <T> (mg cm $^{-2}$) | 363.3 | 659.6 | 364.1 | 394.3. | 255.0 | 165.5 |
| Change in deposits thickness <D> (cm) | 0.2 | 2.5 | 1.34 | 3.3 | 0.0 | 0.0 |
| Net deposits <N> (mg cm $^{-2}$) | 11.2 | 140.2 | 136.5 | 346.5 | 0.0 | 0.0 |
| Removal rate. <R>=<T> - <N> (mg cm $^{-2}$) | 352.1 | 519.4 | 227.6 | 47.8 | 255.0 | 165.5 |
| Removal Coeff. <R>/<T> (%) | 97 | 79 | 63 | 12 | 100 | 100 |

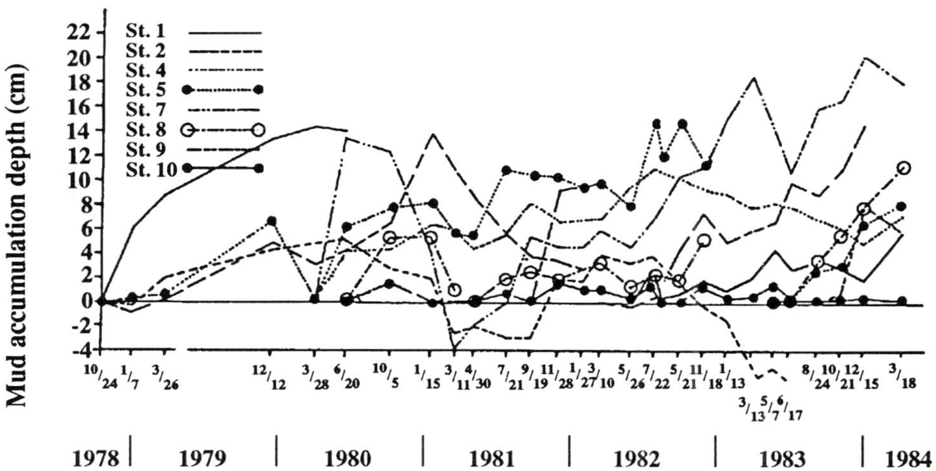


Figure 18.6. Temporal changes in the bottom surface level measured with a deposit thickness measuring plate.

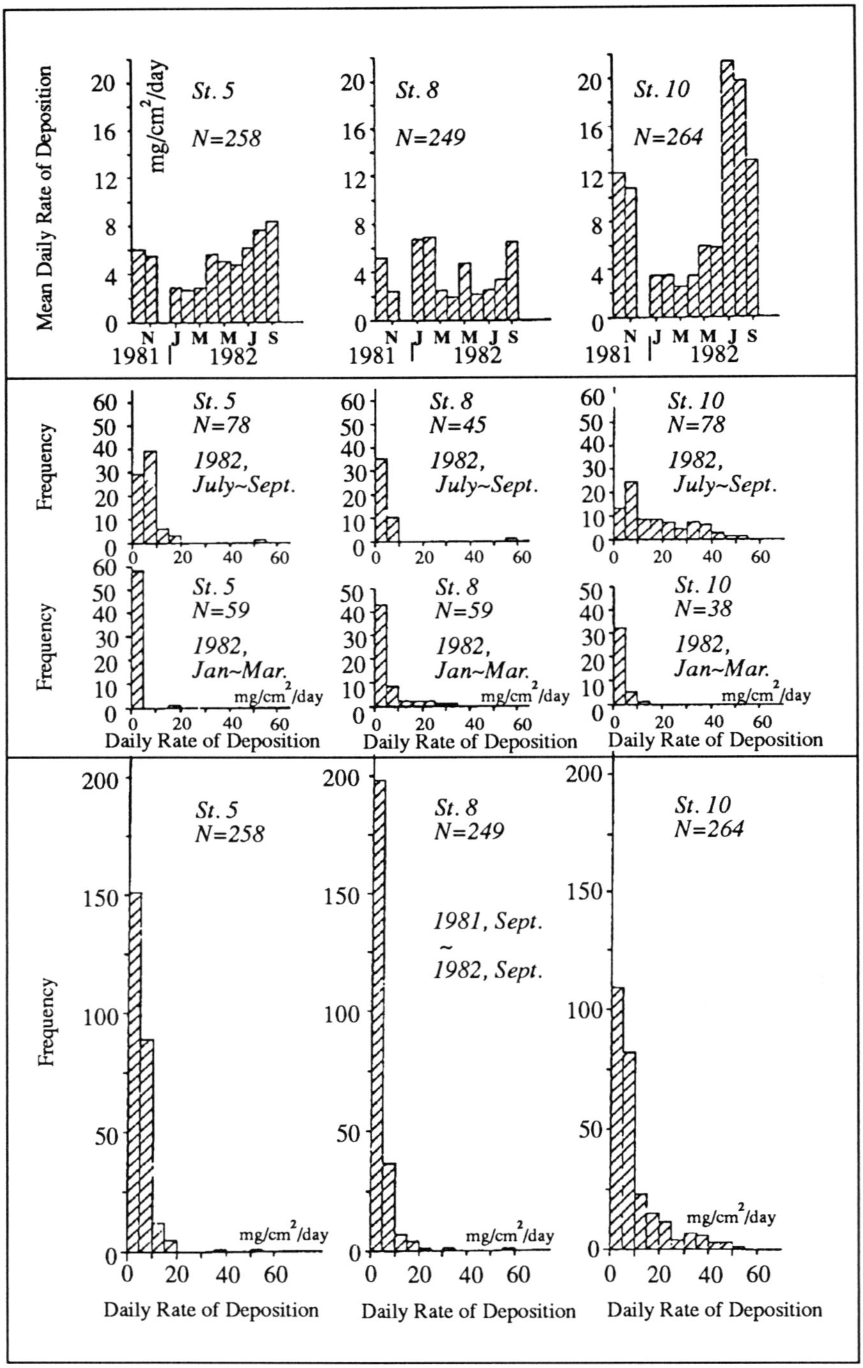


Figure 18.7. Observation results using a daily sediment trap.

## Conclusion

We have developed several instruments and used them for field observations to measure the long-term and short-term variation of depositional processes at the lake bottom. There has been some progress in measuring techniques for verifying bottom deposit thickness in shallow areas near the lake shore. Nevertheless, a number of unsolved problems remain.

In the thermistor deposit meter the sensors are presently set at intervals of 5 mm, but should be at less than 3 mm for more accurate measurement. The present measurement principle of the thermistor deposit meter assumes the temperature of water and deposits to be almost the same. However, temperature changes of the deposits lags behind variations in ambient water temperatures. Under such circumstances, accurate observation requires that the temperatures of the water and deposits be measured independently. Improvement is also needed in the daily sediment traps so that the trapping ratio becomes constant under different hydromechanical conditions. Finally, measurement of the variation of the bottom surface level by deposit thickness measuring plates must be improved so that setting and measurement can be carried out from boats or land without the need for diver assistance.

## References

Edington, D.N. and J.A. Robbins (1976) *Environmental Biogeochemistry*, 2. J. O. Nriague (ed). Ann Arbor, Science, MI., pp 705-729.

Okuda, S., K. Yokoyama, K. Nishi and M. Kumagai, (1986) Study on movement of bottom deposit, Part 1. Method and result of observation, *Annuals, Disa. Prev. Res. Inst. Kyoto Univ.*, 29B-1, 341-353.

Okuda, S. K. Yokoyama, K. Nishi and M. Kumagai, (1986) Observation method on movement of lake bottom deposit, *Res. Rep. from Lake Biwa Res. Inst.*, No 86-A04, 59-64.

Kumagai, M., S. Okuda and K. Yokoyama, (1986) The sediment resuspension in Lake Biwa, *Res Rep. from Lake Biwa Res. Inst.* No 86-A04, 71-82.

# List of Contributors

S. Endoh
Faculty of Education
Shiga University
2-5-1 Hiratsu, Otsu, Shiga, 520
Japan

K. Fukagae
Teral Kyokuto Inc
Yasuda Building
3-5-5 , 101 Iwamoto-cho
Chiyoda-ku
Japan

K. Fukuyama
Faculty of Education,
Mie University
1515 Uehama-cho, Tsu,
Mie, 514
Japan

H. Fushimi
Lake Biwa Research Institute
1-10 Uchidehama
Otsu, Shiga, 520
Japan

Y. Itakura
Faculty of Education,
Shiga University
2-5-1 Hiratsu, Otsu
Shiga, 520
Japan

Y. Kakui
Life Electronics Research Centre
3-11-46 Nakoji Electrotechnical
Laboratory
Amagasaki , Hyogo 661
Japan

K. Kashiwaya
Faculty of Science
Kanazawa University
Kakuma-cho, Kanazawa
Ishikawa 920-11
Japan

M. Kumagai
Lake Biwa Research Institute
1-10 Uchidehama
Otsu, Shiga, 520
Japan

M. Nakanishi
Centre for Ecological Research
Kyoto University
4-1-23 Shimosakamoto,
Otsu, Shiga, 520
Japan

I. Okamoto (Emeritus Professor)
Faculty of Educion
Shiga University
2-5-1 Hiratsu, Otsu, Shiga, 520
Japan

Y. Oonishi
Environmental Research Institute Inc.
4-11-1206,Tsuruno-cho, Kita-ku
Osaka 530
Japan

K. Okubo
Institute of Disaster Prevention Research
Kyoto University
Gokasyo, Uji, Kyoto, 611
Japan

S. Okuda
Faculty of Science
Okayama University of Science
1-1 Ridai-cho
Okayama, Okayama, 700
Japan

Y. Okumura
Osaka Electro-Communication
University
18-8 Hatsumachi
Neyagawa, Osaka, 572
Japan

H. Taishi
Kinki Geo-Engineering Centre Co., Ltd
21-1 Kuwanomoto-cho, Yokoouji, 612
Fushimi-ku, Kyoto, Kyoto
Japan

R. Tsuda
Faculty of Agriculture
Kinki University
3327-204 Naka-machi
Nara, Nara, 631
Japan

A. Yamamoto
Osaka Electro-Communication
University
Hatsu-machi, 18-8
Neyagawa 572, Osaka
Japan

K. Yokoyama
Institute of Disaster Prevention Research
Kyoto University
Gokasyo, Uji, 611
Kyoto
Japan